Optics and Artificial Vision

IOP Series in Emerging Technologies in Optics and Photonics

Series Editor

R Barry Johnson, a Senior Research Professor at Alabama A&M University, has been involved for over 50 years in lens design, optical systems design, electro-optical systems engineering, and photonics. He has been a faculty member at three academic institutions engaged in optics education and research, employed by a number of companies, and provided consulting services.

Dr Johnson is an IOP Fellow, SPIE Fellow and Life Member, OSA Fellow, and was the 1987 President of SPIE. He serves on the editorial board of *Infrared Physics & Technology* and *Advances in Optical Technologies*. Dr Johnson has been awarded many patents, has published numerous papers and several books and book chapters, and was awarded the 2012 OSA/SPIE Joseph W Goodman Book Writing Award for Lens Design Fundamentals (second edition). He is a perennial co-chair of the annual SPIE Current Developments in Lens Design and Optical Engineering Conference.

Foreword

Until the 1960s the field of optics was primarily concentrated in the classical areas of photography, cameras, binoculars, telescopes, spectrometers, colorimeters, radiometers, etc. In the late 1960s optics began to blossom with the advent of new types of infrared detectors, liquid crystal displays (LCDs), light emitting diodes (LEDs), charge coupled devices (CCDs), lasers, holography, fiber optics, new optical materials, advances in optical and mechanical fabrication, new optical design programs, and many more technologies. With the development of the LED, LCD, CCD, and other electro-optical devices, the term 'photonics' came into vogue in the 1980s to describe the science of using light in the development of new technologies and the performance of a myriad of applications. Today optics and photonics are truly pervasive throughout society and new technologies are continuing to emerge. The objective of this series is to provide students, researchers, and those who enjoy self-education with a wide-ranging collection of books that each focus on a relevant topic in the technologies and applications of optics and photonics. These books will provide knowledge to prepare the reader to be better able to participate in these exciting areas now and in the future. The title of this series is *Emerging Technologies in Optics and Photonics* where 'emerging' is taken to mean 'coming into existence', 'coming into maturity', and 'coming into prominence'. IOP Publishing and I hope that you find this series of significant value to you and your career.

Optics and Artificial Vision

Rafael G González-Acuña
Université Laval, Québec, Canada

Héctor A Chaparro-Romo
Independent Research Alcanfor Mz.8 Lt.8, Tultitlán Estado de México, México

Israel Melendez-Montoya
Alestra, Monterrey, México
and
México Industrias Rochin, Culiacán, México

IOP Publishing, Bristol, UK

ISBN 978-0-7503-3707-6 (ebook)
ISBN 978-0-7503-3705-2 (print)
ISBN 978-0-7503-3708-3 (myPrint)
ISBN 978-0-7503-3706-9 (mobi)

DOI 10.1088/978-0-7503-3707-6

Version: 20210901

IOP ebooks

British Library Cataloguing-in-Publication Data: A catalogue record for this book is available from the British Library.

Published by IOP Publishing, wholly owned by The Institute of Physics, London

IOP Publishing, Temple Circus, Temple Way, Bristol, BS1 6HG, UK

US Office: IOP Publishing, Inc., 190 North Independence Mall West, Suite 601, Philadelphia, PA 19106, USA

In memory of Professor Reinhard Klette (1950–2020).

Contents

Preface

In this book we cover the fundamentals of computer vision. Computer vision is a constantly growing branch of research that has attracted attention for its successful applications in industry and engineering.

In this book we study the branches of science that make up computer vision. We study the main algorithms of these branches and also the mathematical concepts behind them.

This book deals with the following topics. In chapter 1 we study what computer vision is and of what it consists. Digital image processing in `Python` is covered in chapter 2. In chapter 3 we study the algorithms of optical flow. In chapter 4 the algorithms of object detection are studied. In chapter 5 keypoint image descriptors for classical artificial vision are described and implemented.

In chapter 6, a quick theoretical and implementational overview of convolutional neural networks for typical vision tasks is seen. In chapter 7, the basics of optical character recognition are viewed as well as basic open source-ready implementations. In chapter 8, we review the historical basis of facial recognition up until the modern deep learning implementations. In chapter 9, we briefly review and provide implementations of four state-of-the-art modern vision cases. We present all the codes of all the algorithms that we study in this book in `Python`. Generally all the codes in `Python` are explained. On the occasions that the explanation is omitted this is because the code comments are explicit and are formulated to explain the code. It is considered a prerequisite that the reader has a basic knowledge of the programming language `Python`.

This introductory book can be of great help for people who want to jump start their introduction into computer vision with a quick overview of the principle components of classical and modern vision techniques. This book is ideal for industry engineers with projects related to computer vision and can be used as a reference for academics, students and researchers in this area.

—The Authors

Acknowledgements

Acknowledgements of Rafael G González-Acuña

This book aims to honour the glory of God. God, ultimate truth of truths, thank you for helping me with this book, for guiding me and illuminating my path.

I would like to thank Mary, *notre dame, reine le monde*, for being a mother, advocate, intermediary between us and the Almighty, and for helping me with this book.

I want to thank my mother Carmen Leticia Acuña Medellín, my father Rogelio González Cantú, my brothers and the rest of the family. I also want to thank my extended family in México City and in Costa Rica.

Héctor A Chaparro-Romo and Israel Melendez-Montoya, my comrades, once again bros—mission accomplished! Héctor you are a person with mature and well-formed thinking from which I can always learn by analysing your thoughts and views. Israel you are a young talent from whom I can always learn and you will surprise us—keep going!

My thanks to Professor R Barry Johnson for your support and trust, and to Ashley Gasque and Robert Trevelyan and the whole IOP team.

I would also like to thank the following people: Professor Julio C Gutiérrez-Vega, Dr Bernardino Barrientos García, Mayra Vargas, Dr Laura Carolina Narro Dra, Anne-Sophie Poulin-Girard, Adriana Mabel Serrano Garza, Jeck Borne, Professor Simon Thibault, Dr Gustavo Media, Roberto Vera, Dr Dora Medina, Dr Alberto Silva and Dr Rafael Torres.

I wrote this book in memory of Professor Reinhard Klette (1950–2020) who was one of my first mentors, someone who gave me excellent advice and encouraged me to follow a career in science. With Professor Reinhard Klette I published my first paper which was on the topic of this book.

I would also like to thank several institutions: the Institute of Physics, Université Laval, Conacyt, Instituto Tecnológico y de Estudios Superiores de Monterrey, Wolfram Research, Centro de Investigaciones en Optica AC, Auckland University of Technology, Institut für Technische Optik Universität Stuttgart, Universidad Abierta y a Distancia de México, Universidad Yachay Tech and Oxford Immune Algorithmics.

Acknowledgements of Héctor A Chaparro-Romo

I want to thank the wider IOP family and am particularly grateful to Professor Dr Barry Johnson, Editor-in-Chief of the series, Ashley Gasque and Robert Trevelyan for their support and instructions.

I thank you, dear readers, for your patience and the decision to read our work, which aims to be an invitation into the world of computational image/vision through accessible computer tools. Here we present a brief view of the potential depth of these technological toolkits.

My family, friends and teachers, thank you for always being with me and supporting my adventures, we continue to enjoy science thoroughly.

Rafael and Israel, thank you for your commitment to achieve the goal proposed for this work, congratulations!

Acknowledgements of Israel Melendez-Montoya

I primarily thank my fellow authors Rafael and Héctor for their patience and the opportunity to work with two of the greatest minds in geometrical optics that I have ever known. They are not only committed to the development of optical science but are also people who can prove that there is always a problem waiting to be solved.

I would like to thank the Institute of Physics, specifically Professor Dr Barry Johnson, Ashley Gasque and Robert Trevelyan, for all their support. Without them this work would not have been possible.

I would also like to thank my family for their patience and support, always.

I thank Marco de los Santos and the people at AxtelLabs: Rómulo, Hernán, Paco, Uziel, Jesús and Audi. I would not be working in artificial vision if it were not for them.

I would like to thank the professors at the Physics Department at Tecnológico de Monterrey, specifically my advisor Dr Dorilián, for believing in me academically and teaching me how good science is done. I would like to thank specifically the people who were there for the academic and non-academic times while I was at Tecnológico de Monterrey: Benjas, Yepiz, Salvador Elías, Blas, Israel de León, Ebrahim Karimi, Luis Sanchéz-Soto, Ale de Luna, Diana, Job, Luis Garza, Rojas, Uriel, Karla, Mateusz, and the *nanoamiguitos*: Manuel, Hugo and Matías. I am grateful for all their teachings.

I thank the hard-working professors at UANL-FCFM, Tlahuice, Gámez, Conchis, Edgar, Polito, Francisco, Tenorio, Abigail, Adriana, and others who I may not have met, for giving students something to believe in and trusting us in their teaching process, this book would not be here without the basis of mathematics and physics that professors provide. I thank also my graduate generation at FCFM: Omar, Neri, Daniel, Norman, Saúl, Simon, Marco, Rodrigo and Hiram for all of their academic conversations, studying and competitiveness. I strongly believe that the key role in our success was played by our effective communication.

I thank the UASLP graduate physics department for accepting third year students and letting us see the formalism and next-level hard work necessary to continue to study graduate-level physics.

I would like to thank my friends the Feferis, Rafa, Eliel, Mike, Mora, Hohomer, Rojo, Feri, Joel, Mawa and Ferras, for their company and the ongoing adventures I never thought I would have experienced a few years ago, as well as for giving me life advice into the future. For Eugenia, thank you for your space and time.

And finally, special thanks to the reader who will take the time to read the contents of this book.

Author biographies

Rafael G González-Acuña

Rafael G González-Acuña studied industrial physics engineering at the Tecnológico de Monterrey and received his master's degree in optomechatronics from the Centro de Investigaciones en Óptica, AC. He is currently studying for his PhD at the Tecnológico de Monterrey. His doctoral thesis focuses on the design of spherical aberration free lenses. He is the co-author of the solution to the problem of designing bi-aspheric singlet lenses free of spherical aberration. He is the co-author of the books *Analytical Lens Design* and *Stigmatic Optics*.

Héctor A Chaparro-Romo

Héctor A Chaparro-Romo obtained his bachelor's degree in electronic engineering at the Universidad Autónoma Metropolitana. He is the co-author of the solution to the problem of spherical aberration in lens design and is also the co-author of the solution to the design of a mirror whose property is to deliver a flat wavefront given any wavefront. He is the co-author of the books *Analytical Lens Design* and *Stigmatic Optics*.

Israel Melendez-Montoya

Israel Melendez-Montoya studied physics at the Universidad Autónoma de Nuevo León and optics at Tecnológico de Monterrey and works for Alestra in the Technology Innovation department as a Data Scientist. He has four years of experience in image data processing and uses artificial intelligence and neural network paradigms. His favorite Python (and C) libraries are Tensorflow, OpenCV and PyTorch using the `Python` language. Many of his projects have been developed over open source code. His main interests include studying deep learning vision techniques, going to the gym and dogs.

Rafael G González-Acuña, Héctor A Chaparro-Romo and Israel Melendez-Montoya

Chapter 1

Optics, sensors and images

In this chapter we introduce the basic notions behind artificial vision. We first review the role of optics and image formation in visual phenomena (Born and Wolf 2013, Chartier 2005, King *et al* 2000, Kingslake and Johnson 2009). Then we study visual phenomena and the human eye and other optical systems relevant to artificial vision. Finally, we explore the hardware and software concepts and technology behind artificial vision.

1.1 Introduction

Before we start to address all the concepts, ideas and information in this chapter, it will be useful to set the objective of this chapter. The goal is to relate all the disciplines that relate to artificial vision.

In section 1.2 we will start with optics, as for the authors this is the most crucial topic in vision and artificial vision. Then in section 1.3 we briefly study vision and the human eye, which is essential to do it since the images captured by our eyes are our first reference for visual phenomena. Optical systems can be seen as an upgrade to our vision and they are discussed in section 1.4. Cameras can improve our vision, but they are also a fundamental part of the goal of generating what we can call artificial vision. Thus in section 1.5 we discuss cameras and in sections 1.6 and 1.7 we focus on the primary sensors found in digital cameras. Once we understand the hardware behind artificial vision, we need to define what artificial vision is and also the algorithmic disciplines that support it. This discussion takes place in section 1.8. Finally, in section 1.9 we talk a little bit about `Python`, which is the programming language in which all the codes in this book are written. This book assumes that the reader has a basic knowledge of `Python` (Chaves 2017, Gross *et al* 2005).

doi:10.1088/978-0-7503-3707-6ch1

1.2 Optics and images

Optics is the branch of physics that studies light and its nature. Thus it explores the behaviour and properties of light. Optics includes the interactions of light with matter and other light sources. For this, optical scientists usually use instruments to detect, manipulate and study the behaviour of light.

One of these instruments is the human eye, which will be discussed in the next section. But before we continue, it is essential to remember that optics is a subsection of the study of electromagnetic radiation. Optics studies only visible light. Note that the word visual somehow implies what is visible to the human eye. Although this definition seems quite ambiguous, formally optics generally describes the behaviour of visible light, ultraviolet radiation and infrared radiation. Other forms of electromagnetic radiation, such as x-rays, microwaves and radio waves, are not directly studied by optics although they share similar properties.

Another important observation is that optics studies visible light under the electrodynamics description or classical physics description. Thus optics does not deal with the quantum phenomena of light and, instead, photonics is the discipline that studies the quantum phenomena of light. However, for most applications of artificial vision quantum phenomena can be ignored. Hence we are going to focus on geometric optics, which is a simplified model of the behaviour of light.

Geometric optics treats light as a collection of rays that travel in a straight line and deflect when they pass through or are reflected off surfaces. From geometric optics we can deduce the concept of an image.

An optical image is a figure formed by the set of points where the rays that come from point sources of the object converge after their interaction with the optical system. The image can be of two types: real or virtual.

A real image is one that is formed when, after passing through the optical system, the light rays converge. We cannot perceive this image directly with our sense of sight, but it can be recorded by placing a screen in the place where the rays converge.

A virtual image is one that is formed when, after passing through the optical system, the rays diverge. To our sense of sight the rays seem to come from a point that they have not really passed through. The image is perceived in the place where the prolongations of these divergent rays converge. This is the case of an image formed by a flat mirror.

The sense of sight or vision is the capability to detect images. Thus, as we mentioned above, optical instruments help to study the behaviour of light, but also to detect it in the form of images. The first instrument used to detect the light of images used by humankind is their eyes, so we are going to discuss the eye in the next section.

1.3 Vision

The sense of sight or vision is possible thanks to a receptor organ, the eye, which receives light impressions and transforms them into electrical signals that are transmitted to the brain through optical pathways. The eye is a paired organ located in the orbital cavity. It is protected by the eyelids and by the secretions of the lacrimal gland.

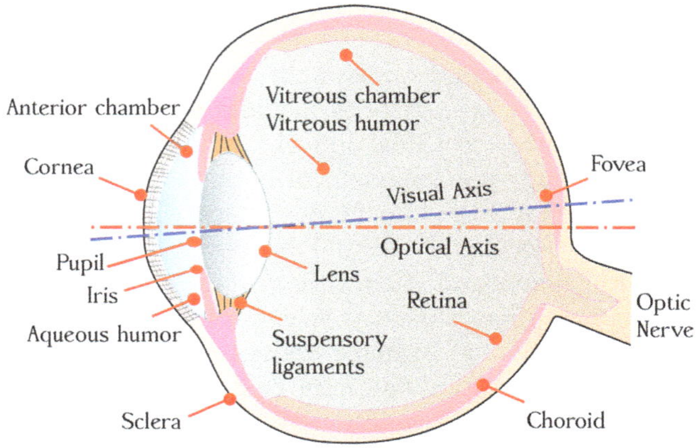

Figure 1.1. A diagram of the human eye.

It can move in all directions thanks to the extrinsic muscles of the eyeball. The essential property that makes vision possible is photosensitivity. Photosensitivity takes place in specialized receptor cells that contain chemical substances that are capable of absorbing light to produce a photochemical change, see figure 1.1 (Kaufman 1974).

When light enters the eye it passes through the cornea, the pupil and the lens to reach the retina, where the electromagnetic energy of the light is converted into nerve impulses that are sent through the optic nerve to the brain for processing by the visual cortex. In the brain the complicated process of visual perception takes place, thanks to which we can perceive the shape of objects, identify distances, and detect colours and movement. The retina is one of the most critical regions of the eye and contains specialized cells called rods and cones that are sensitive to light.

The eye is the organ responsible for receiving visual stimuli and it has a highly specialized structure. In humans, the eye has three layers, which from outside to inside are as follows.

The *external fibrous tunic* is made up of two regions, the sclera and the cornea.

1. The *sclera* is white and opaque, made of a type of collagen fibre interspersed with elastic fibres. It is avascular and provides protection and stability to the internal structures. The sclera covers most of the eyeball, except for a small anterior region.
2. The *cornea* is transparent and avascular but highly innervated. The cornea is slightly thicker than the sclera.

The *middle vascular tunic (uvea)* is made up of three regions: the choroid, the ciliary body and the iris.

1. The *choroid* is the pigmented posterior portion of the medial vascular tunic, which loosely joins the sclera and is separated from the lens by Bruch's membrane.
2. The *ciliary body* is a cuneiform prolongation which projects towards the lens and is located in the lumen of the eye between the iris (anterior) and the vitreous humour (posterior).

3. The *iris* is the pigmented anterior extension of the choroid, whose function is to regulate the entry of light into the eye through the contraction or distention of the pupil.

Finally, we have the *retina* or *neural tunica*. It is the light-sensitive portion of the eye where specialized cells are found. It is composed of ten layers which, from the outside to the inside, are called the pigmented epithelium, cone and rod (receptor) layer, outer limiting membrane, outer nuclear layer, outer plexiform layer, inner nuclear layer, inner plexiform layer, cell layer ganglion cells, optic nerve fibre layer and internal limiting membrane.

It is important to note that colour is not a property of light or reflective objects, but rather a cerebral sensation. Humans see colours as a result of the interaction of light in the eye through the ocular structure of cones, which detect the energy of photons, transmitting the sensation to the brain. The perception of colours is subjective and depends on the attributes that the brain assigns to specific wavelengths. In this way a wavelength of 560 nm is defined as red. However, in reality, red or any other colour do not exist, only electromagnetic radiation with a given wavelength is real (Snowden *et al* 2012, Wagemans *et al* 2005).

1.4 Optical instruments and optical design

Although human vision and the eye are fascinating objects of study, in practice they are limited. For example, the smallest object that can be seen with the naked eye cannot be less than 0.1 mm in size, the thickness of some fibres or a human hair. Therefore humans have developed all kinds of optical instruments to improve their vision. Examples of these instruments are cameras, ophthalmic glasses, microscopes, telescopes, etc (González-Acuña and Chaparro-Romo 2020, González-Acuña *et al* 2020, O'Shea 1985, Smith *et al* 2007, Velzel 2014).

The optical mechanisms possessed by the human eye are those used in optical instruments. By altering the radii of curvature and refractions, the focal length is manipulated and the light rays are focused, which enlarges the objects. Microscopes, magnifying glasses, binoculars and telescopes are based on this simple principle, see figure 1.2.

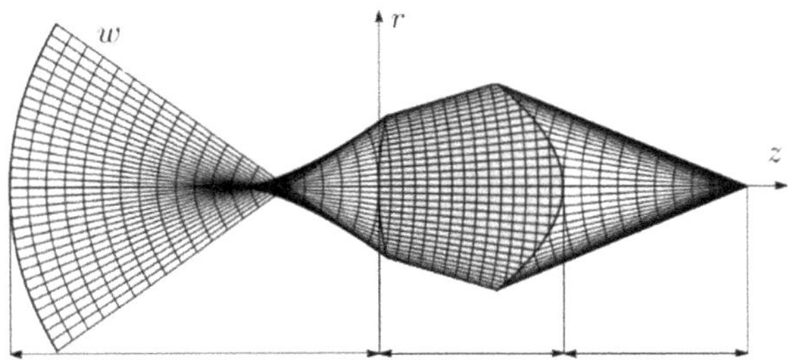

Figure 1.2. Optical design scheme.

All this development set the basis for the hardware of artificial vision, which are the instruments used to detect an image in an artificial way. The most common device for this task is the digital camera.

1.5 Cameras

A digital camera is a photographic camera that, instead of capturing and storing photos on a chemical film like photographic film cameras, uses digital photography to generate and store images (Gernheim and Gernheim 1969).

The camera's sensor limits the resolution of a digital photographic camera. Usually, a CCD or a CMOS sensor responds to light signals, replacing the work of film in traditional photography. The sensor is made up of millions of *cubes* that are charged in response to light. Generally, these cubes only respond to a limited range of light wavelengths due to a colour filter on each one. Each of these cubes is called a *pixel*. A mosaic of pixels is implemented to generated an RGB image where all three images per pixel are to represent a full colour (Gregorian 1999, Holst 1998).

An RGB image is an image formed by three different channels, one in red, one in green and one in blue. The red channel has the information of the pixels that captured red light. An equivalent process happens for the green and blue channels (North 2005).

Thus an RGB image I_{RGB} can be defined by a matrix $N \times M \times 3$, where N is the number of pixels wide and M is the number of pixels high. Thus,

$$I_{RGB} = [I_R, I_G, I_B] \tag{1.1}$$

where, I_R, I_G and I_B, are the red, blue and green channels respectively.

When an image is formed by only one channel it is usually called a greyscale image. Greyscale is a scale used in digital images in which the value of each pixel has an amount equivalent to a gradation of grey. Rendered images of this type are composed of shades of grey.

The resulting number of pixels in the image determines its size. For example, an image $N = 640$ pixels wide by $M = 480$ pixels high will have $N \times M = 307\,200$ pixels, or approximately 307 kilopixels. An image $N = 3872$ pixels tall by $M = 2592$ pixels wide will be $N \times M = 10\,036\,224$ pixels, or approximately 10 megapixels.

1.6 CCD sensor

The pixels of the CCD register graduations of the three primary colours red, green and blue, as mentioned earlier. A set three pixels, one for each colour, forms a group of photoelectric cells capable of capturing any colour in the image. To achieve colour separation, most CCD cameras use a Bayer mask that provides a raster for each set of four pixels so that one pixel registers red light, another blue light and two pixels are reserved for green light, see figure 1.3. The final result includes information about the luminosity in each pixel but with a colour resolution lower than the illumination resolution (Gregorian 1999, Holst 1998).

CCD detectors, like photovoltaic cells, are based on the photoelectric effect, the spontaneous conversion of received light into electrical current that occurs in some

Figure 1.3. Bayer filter scheme.

materials. The sensitivity of the CCD detector depends on the quantum efficiency of the chip—the number of photons that must hit each sensor to produce an electrical current.

1.7 CMOS sensor

Since the CCD sensor is based on the photoelectric effect, it is made up of numerous light collectors, called photosites, one for each pixel, which produce an electric current that varies depending on the intensity of light received. In a CMOS, unlike a CCD, an electrical signal amplifier is incorporated in each photosite and it is common to include the digital converter on the chip itself. While in a CCD the electrical signal produced by each photosite has to be sent to the outside, where the electrical signal is amplified to a computer, the advantage of a CMOS is that the electronics can read the signal directly from each pixel. As in CCDs, in digital cameras with CMOS sensors Bayer filters are also implemented to generate RGB images (Gregorian 1999, Holst 1998, Maloberti 2006, Waltham 2013).

1.8 `Python` as a program language for this book

`Python` is an interpreted programming language whose philosophy emphasizes the readability of its code. It is a multiparadigm programming language, as it supports object-oriented, imperative programming, and, to a lesser extent, functional programming. It is an interpreted, dynamic and multiplatform language.

`Python` is of interest to us because it is free and it has several libraries of functions for image analysis and machine learning. Also, the documentation of each library is growing, as well as its community (Lutz 2001, Van Rossum and Drake 1995).

As a prerequisite for this book the reader must be familiar with `Python`. However, all the algorithms implemented will be deduced step by step, so `Python` is not required in the explanations. `Python 3` will be needed in order to run the codes of each chapter. At the beginning of each code the names of the relevant libraries of `Python` will be given.

1.9 Artificial vision and computer vision

Now that we have outlined the basis of the disciplines that made artificial vision possible—optics, the sense of sight and the hardware—it is the proper time to define what artificial vision is. Artificial vision, also known as computer vision, is a scientific discipline that includes methods to acquire, process, analyse and understand images of the real world to produce numerical or symbolic information that can be processed by a computer (Fischer *et al* 2000, Forsyth and Ponce 2002).

Thus, just as we humans use our eyes and brains to understand the world around us, computer vision tries to produce the same effect such that computers can perceive and understand an image or sequence of images and act as appropriate in a given situation. This understanding is achieved thanks to different fields such as geometry, statistics, optics, etc. Geometry and statistics are commonly used in artificial vision to extract information from the images. Optics are used in computer vision to understand the formation of the images that will be perceived by the computer (Kropatsch *et al* 2012, Pedregosa 2011, Raschka 2015, Rosenfeld 1988).

Data acquisition is achieved by various means such as image sequences, viewed from multiple video cameras, or multidimensional data from a medical scanner. The principles behind the CCD and CMOS sensors can be seen as part of the data acquisition.

Since artificial vision is a discipline composed of several other disciplines, there are several similar fields. The fields most closely related to computer vision are image processing and image analysis. There is significant overlap in the range of techniques and applications that they cover. The following list presents several related fields, highlighting the differences and similarities between them.

1. *Artificial vision* or *computer vision* is the process of applying a range of technologies and methods to provide automatic image-based inspection, process control and robotic guidance in industrial applications. Computer vision tends to focus on applications, primarily manufacturing; for example, vision-based robots and systems for vision-based inspection or measurement. This implies that image sensor technologies and control theory are often integrated with image data processing to control a robot, and that real-time processing is achieved by efficient hardware and software implementations (Granlund and Knutsson 2013, Klette 2014, Szeliski 2010).
2. *Image processing*, as the name suggests, is the process of modifying an image. Common methods of transforming images are threshing, segmentation, edge detection, dilatation and contraction (Gonzalez *et al* 2004, Klette 2014, Schalkoff 1989, Szeliski 2010).
3. *Imaging* is a field primarily focused on the image production process. The type of optical instrument used to generate the images and the kind of sensor that detects the images are the most common areas of focus in imaging (Klette 2014, Szeliski 2010).

4. *Pattern recognition* is a field that uses various methods to extract information from signals in general, mainly based on statistical approaches and artificial neural networks. A significant part of this field is devoted to applying these methods to image data. When pattern recognition is applied in images it can be seen as a tool of artificial vision (Klette 2014, Szeliski 2010).
5. *Machine learning* techniques aim to differentiate patterns automatically using mathematical algorithms. These techniques are commonly used to classify images and to make decisions. Two main types of methods can be distinguished: supervised and unsupervised. Like pattern recognition, machine learning can be applied to differentiate patterns in image data as a part of the application of artificial vision. In supervised learning the computer is trained by providing previously labelled patterns, so that the algorithm used must find the boundaries that separate the different possible types of patterns. In unsupervised learning, the computer is trained with patterns that have not been previously classified, and it is the computer itself that must group the different patterns into different classes. *K*-means and some neural networks are part of the unsupervised learning group. Both techniques are widely used in artificial vision, particularly in image classification and segmentation (Klette 2014, Szeliski 2010).

Thus all the fields and disciplines presented in the above list are parts or tools of an artificial vision system. An artificial vision system is a system made of hardware (sensors, camera, computer) and software (algorithms), such that it has the sense of sight or vision for a particular task. We use the `OpenCV` library for computer vision tasks (Oliphant 2007).

1.10 End notes

In this chapter, we review the essential concepts and disciplines of computer vision. We also define artificial vision and the branches related to it. If you are looking to go deeper into the use and scope of these technologies, it is highly recommended to use the information in the manuals and official sites in the references.

References

Born M and Wolf E 2013 *Principles of Optics: Electromagnetic Theory of Propagation, Interference and Diffraction of Light* (Amsterdam: Elsevier)

Chartier G 2005 *Introduction to Optics* (Berlin: Springer)

Chaves J 2017 *Introduction to Nonimaging Optics* (Boca Raton, FL: CRC Press)

Fischer R E, Tadic-Galeb B and Yoder P R 2000 *Optical System Design* (New York: McGraw-Hill)

Forsyth D A and Ponce J 2002 *Computer Vision: A Modern Approach* (Englewood Cliffs, NJ: Prentice Hall)

Gernheim H and Gernheim A 1969 *The History of Photography from the Camera Obscure to the Beginning of the Modern Era* (New York: Thames & Hudson)

Gonzalez R C, Woods R E and Eddins S L 2004 *Digital Image Processing Using MATLAB* (New Delhi: Pearson Education India)

González-Acuña R G and Chaparro-Romo H A 2020 *Stigmatic Optics* (Bristol: Institute of Physics Publishing)
González-Acuña R G, Chaparro-Romo H A and Gutiérrez-Vega J C 2020 *Analytical Lens Design* (Bristol: Institute of Physics Publishing)
Granlund G H and Knutsson H 2013 *Signal Processing for Computer Vision* (Berlin: Springer)
Gregorian R 1999 *Introduction to CMOS op-amps and Comparators* (New York: Wiley)
Gross H, Singer W, Totzeck M, Blechinger F and Achtner B 2005 *Handbook of Optical Systems* vol 1 (New York: Wiley)
Holst G C 1998 *CCD Arrays, Cameras, and Displays* (Portal, OR: SPIE)
Kaufman L 1974 *Sight and Mind: An Introduction to Visual Perception* (New York: Oxford Book Company)
King T A, Graham-Smith F and Smith Sir F G 2000 *Optics and Photonics: An Introduction* (New York: Wiley)
Kingslake R and Johnson R B 2009 *Lens Design Fundamentals* (New York: Academic)
Klette R 2014 *Concise Computer Vision* (Berlin: Springer)
Kropatsch W, Klette R, Solina F and Albrecht R 2012 *Theoretical Foundations of Computer Vision* vol 11 (Berlin: Springer)
Lutz M 2001 *Programming Python* (Sebastopol, CA: O'Reilly)
Malacara-Hernández D and Malacara-Hernández Z 2016 *Handbook of Optical Design* (Boca Raton, FL: CRC Press)
Maloberti F 2006 *Analog Design for CMOS VLSI Systems* vol 646 (Berlin: Springer)
North M 2005 *Camera Works: Photography and the Twentieth-Century World* (Oxford: Oxford University Press)
Oliphant T E 2007 Python for scientific computing *Comput. Sci. Eng.* **9** 10–20
O'Shea D C 1985 *Elements of Modern Optical Design* (New York: Wiley)
Pedregosa F *et al* 2011 SciKit-learn: machine learning in Python *J. Mach. Learn. Res.* **12** 2825–30
Raschka S 2015 *Python Machine Learning* (Birmingham: Packt)
Rosenfeld A 1988 Computer vision: basic principles *Proc. IEEE* **76** 863–8
Schalkoff R J 1989 *Digital Image Processing and Computer Vision* vol 286 (New York: Wiley)
Smith F G, King T A and Wilkins D 2007 *Optics and Photonics: An Introduction* (New York: Wiley)
Snowden R, Snowden R J, Thompson P and Troscianko T 2012 *Basic Vision: An Introduction to Visual Perception* (Oxford: Oxford University Press)
Szeliski R 2010 *Computer Vision: Algorithms and Applications* (Berlin: Springer)
Van Rossum G and Drake F L Jr 1995 *Python Tutorial* vol 620 (Amsterdam: Centrum voor Wiskunde en Informatica)
Velzel C H F 2014 *A Course in Lens Design* vol 183 (Berlin: Springer)
Wagemans J, Wichmann F A and Op de Beeck H 2005 Visual perception I: basic principles *Handbook of Cognition* (London: Sage) pp 3–47
Waltham N 2013 CCD and CMOS sensors *Observing Photons in Space* (Berlin: Springer) pp 423–42

IOP Publishing

Optics and Artificial Vision

Rafael G González-Acuña, Héctor A Chaparro-Romo and Israel Melendez-Montoya

Chapter 2

Introduction to computer vision

In this chapter we review the basis of computer vision. Algorithms to load, display, write and extract information from the images in `Python` are presented. We also discuss the concepts of morphological operations, colour spaces, thresholding, gradients and histograms.

2.1 Loading and saving images

Although a basic knowledge of `Python` is a prerequisite, we are going to start with the basics of `OpenCV`. `OpenCV` is a library of programming functions mainly aimed at real-time computer vision. There are bindings in `Python`, `Java` and `MATLAB/OCTAVE`. In `Python` we call `OpenCV` as we would any other library (López *et al* 2016).

The `Python` code in listing 2.1 loads an image from the disk, displays the image and writes the image (see figure 2.1 for the output).

The command `import ocv` loads `OpenCV`, the abbreviation of `OpenCV` being `ocv`. The image is loaded with the command `ocv.imread`. The command `ocv.imread` has as its argument the path where is the image is located. In this case the image is located at "`some/path/image.jpg`" and the variable to which the image is assigned is called `image`. The command `ocv.imshow` receives two arguments, the caption and the image. In this example the caption is `"Image"` and the image is `image`. `ocv.imshow` displays the image and `ocv.imwrite` writes the `image` under the name "`newimage.jpg`". (Lutz 2001, Oliphant 2007, Raschka 2015, Van Rossum and Drake 1995)

The reader is invited to load any image from his/her computer, display it and write it under a different name.


```python
import cv2 as ocv
image = ocv.imread("image.jpg")
ocv.imshow("Image",image)
ocv.imwrite("newImage.jpg",image)
```

Listing 2.1 Load and save image.

Original Image Processed New Image

Figure 2.1. Output of listing 2.1: load and save image. Credit: Highsmith, Carol M., 1946 - United States Library of Congress's Prints and Photographs Division under the digital ID highsm.09920.

2.2 Image basics

Let us talk about some features of images. Images in `Python` are just `NumPy` arrays. The reader should become familiar with what a `NumPy` array is. Every element of the array has a numerical value from 0 to 255. The elements are also called pixels. 0 means black, while 255 is white. Everything in between is a greyscale gradient. If the image is RGB, like those we discussed in the previous chapter, then there are three channels, one for red, one for green and one for blue. The elements of each channel also run from 0 to 255, where black is (0,0,0), white is (255,255,255), red is (255,0,0), green is (0,255,0) and blue is (0,0,255). Every other possible combination is a colour inside the RGB colour scale (Chityala and Pudipeddi 2020, Schalkoff 1989, Szeliski 2010, Tambe et al 2013).

To show the basics of the image, let us study the `Python` code in listing 2.2 (see figure 2.2 for the output).

Since the images are arrays, the pixels are indexed within a coordinate system[1].

The point at (0, 0) corresponds to the top-left pixel in our image, while the point `(h,w)` corresponds to the bottom right corner. We can manipulate each pixel as in line 16.

We can obtain a subimage from the original image such as `topleft`, which is the top-left corner of the original image. We can also manipulate the image by region, as in the image displayed with the caption "`Image with black Top - Left Corner`", where we previously turned the top-left corner black.

[1] Thus all the manipulations that can be done with arrays in `Python` can also be done with images.

```python

import cv2 as ocv # load opencv


image = ocv.imread ( "image.jpg" )

# load the image
(h,w) = image.shape[:2]
ocv.imshow ( " Original " ,image)

# Taking the top - left pixel can be found at (0,0)
(b,g,r) = image[0,0]
print("Pixel at (0,0) - Red:{r}, Green:{g}, Blue:{b}".format(r=r,g=g,b=b) )

# Taking the down - right pixel can be found at (h-1, 0)
(b,g,r) = image[h-1,0]
print("Pixel at ({h},0) - Red:{r}, Green:{g}, Blue:{b}".format(h=h,r=r,g=g,b=b) )


# Change the value of the pixel at (0 , 0) turn it to white
image[0,0] = (255,255,255)
(b,g,r) = image[0,0]
print("Pixel at (0,0) - Red:{r}, Green:{g}, Blue:{b}".format(r=r,g=g,b=b) )

#Computing the centre of the image
(centre_X,centre_Y) = ( w // 2 , h // 2)
# taking the topleft of the original imagew as a new image
topleft = image [0:centre_Y , 0:centre_X]
ocv.imshow ("Top-Left Corner" , topleft)

# turining the top - left corner to black
image[0:centre_Y,0:centre_Y] = (0, 0,0)

# Show our updated image
ocv.imshow("Image with black Top - Left Corner ",image)
ocv.waitKey (0)
```

Listing 2.2 Image basics.

Original Image | Processed New Image

Figure 2.2. Output of listing 2.2: image basics. Credit: Highsmith, Carol M., 1946 - United States Library of Congress's Prints and Photographs Division under the digital ID highsm.09920.

2.3 Colour spaces

There are several colour spaces in images. We have discussed greyscale images and RGB images. In OpenCV there is also BGR, which is similar to RGB but with the blue and red channels inverted (Gonzalez *et al* 2004, Granlund and Knutsson 2013, Klette 2014).

The other colour spaces in OpenCV include the Lab colour spaces. Lab is the short name for two different colour spaces. The best known is CIELAB or L*a*b* and the other is Hunter Lab or L,a,b. Lab is an informal abbreviation and can be confused with one or another colour space. These two colour spaces are related in intention and purpose, but they are different.

The purpose of both spaces is to produce a colour space that is more *perceptually linear* than other colour spaces. Perceptually linear means that a change of the same amount in a colour value should produce a difference of almost the same visual importance. This can improve tone reproduction when colours are stored at limited precision values. Both Lab spaces are related to the white-point of the *XYZ* data from which they were converted. Lab values do not define absolute colours unless a white-point is specified (Forsyth and Ponce 2002).

HSL (hue, saturation, lightness) and HSV (hue, saturation, value), also known as HSB (hue, saturation, brightness), are the other colour spaces. The purpose of these spaces is for the graphics to align with the sensitivity of human vision and perception of colour attributes more closely. In these models the colours of each hue are organized in a radial part, around a middle axis of neutral colours which goes from black at the base to white at the head.

The HSV colour-space models the way paints of different colours mix together, with the saturation dimension matching several tints of brightly coloured paint and the value dimension agreeing with the composition of those paints with differing quantities of black or white.

In most cases we are going to use images in RGB or BGR.

2.4 Basic image processing

In this section we are going to study the basic image processing algorithms.

2.4.1 Translation

Translation is the shifting of an image along the *x*- and *y*-axes. Translation can be achieved simply by multiplying the image by the following matrix:

$$M = \begin{bmatrix} 1 & 0 & t_x \\ 0 & 1 & t_y \end{bmatrix}, \tag{2.1}$$

where t_x and t_y are the translations along the *x*- and *y*-axes, respectively. Let us shift some images and let us do it in Python (listing 2.3 and figure 2.3).

Note that we use numpy to declare the transition matrix M. Also it is important to see that if tx is positive the image is translated to the right and if it is negative the image is translated to the left, and if ty is positive the image is shifted up and if it is negative the image is shifted down.

```python
import numpy as np # load numpy
import cv2 as ocv # load opencv

image = ocv.imread("image.jpg")
ocv.imshow(" Original ", image)

# translation values
tx =20;
ty =30;

# translation matrix
M = np.float32([[1,0,tx],[0,1,ty]])

# translated image
shifted = ocv.warpAffine (image,
            M ,(image.shape[1], image.shape[0]) )
ocv.imshow("Shifted Down and Right ", shifted )

# translation values
tx = -20;
ty = -30;

# translation matrix
M = np.float32([[1,0,tx],[0,1,ty]])

# translated image
shifted = ocv.warpAffine(image, M,(image.shape[1], image.shape [0]))
ocv.imshow(" Shifted Up and Left ", shifted )
ocv.waitKey(0)
```

Listing 2.3 Translation operation.

Figure 2.3. Output of listing 2.3: translation operation. Credit: Highsmith, Carol M., 1946 - United States Library of Congress's Prints and Photographs Division under the digital ID highsm.09920.

The code in listing 2.4 is the function of translation (see figure 2.4 for the output).

2.4.2 Rotation

Rotation is the shifting of an image along the angle θ. Rotation, like translation, is achieved through a matrix multiplication. The rotation matrix is

```
import numpy as np
import cv2 as ocv

image = ocv.imread("image.jpg")
ocv.imshow(" Original ", image)

def translate (image, tx, ty) :
     M = np.float32([[1,0,tx],[0,1 ,ty]])
    shifted = ocv.warpAffine (image, M,(image.shape[1],
        image.shape[0]))
    return shifted

ocv.imshow(" Using translate function ", translate (image
   ,20,30))
ocv.waitKey(0)
```

Listing 2.4 Application of the translation operation.

Original Image   Processed Image

Figure 2.4. Output of listing 2.4: application of the translation operation. Credit: Highsmith, Carol M., 1946 - United States Library of Congress's Prints and Photographs Division under the digital ID highsm.09920.

$$M = \begin{bmatrix} \cos\theta & -\sin\theta \\ \sin\theta & \cos\theta \end{bmatrix}. \tag{2.2}$$

Rotations can be computed with `Python`, using the code in listing 2.5 (see figure 2.5 for the output).

We compute the rotation matrix with the function `cv2.getRotationMatrix2D`, which is fed with the angle of rotation `theta`, the scale `s` and the centre of the image.

The code in listing 2.6 is a function that computes the rotation over a matrix (see figure 2.6 for the output).

2.5 Resizing images

In `Python` we can resize an image with a defined aspect ratio. The aspect ratio is the width of the image divided by its height. Thus the `Python` algorithm in listing 2.7 resizes images (see figure 2.7 for the output).

`dim` is the dimension of the image. The input values `width` and `height` are the amplifications that the image will undergo. Thus if both are `None`, the code returns the same image. If the `width` is `None`, the resize occurs on the height. If the `height` is `None`, the resize occurs on the width of the image. Depending on the specific case

```python
import numpy as np # load numpy
import cv2 as ocv  # load opencv

image = ocv.imread("image.jpg")
ocv.imshow( " Original", image)

# load the image

# computing the dimentions of the image and its centre
(h,w) = image.shape[:2]
(centre_X,centre_Y) = (w //2, h//2)

# rotation angle and scale for rotation Matrix

theta = 45;   s =1;
M = ocv.getRotationMatrix2D((centre_X , centre_Y), theta ,
    s )
rotated = ocv.warpAffine(image, M, (w,h))
ocv.imshow("Rotated by 45 Degrees ",rotated)

theta = -60; s =1;
M = ocv.getRotationMatrix2D((centre_X , centre_Y ),theta ,
    s )
rotated = ocv.warpAffine(image ,M ,(w,h) )
ocv.imshow( "Rotated by -60 Degrees" , rotated)
ocv.waitKey(0)
```

Listing 2.5 Rotation operation.

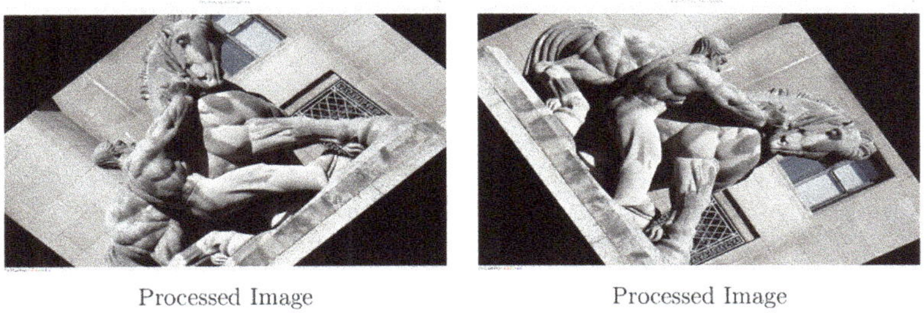

Figure 2.5. Output of listing 2.5: rotation operation. Credit: Highsmith, Carol M., 1946 - United States Library of Congress's Prints and Photographs Division under the digital ID highsm.09920.

the aspect ratio `r` and the `dim` are computed. Finally, focusing on `iter`, this is the type of interpolation in which the resize will take place. To see the different types of interpolations read the `OpenCV` manual.

2.5.1 Flipping

`OpenCV` can also compute rotation on the `x` or `y` axis. The code in listing 2.8 performs this kind of rotation, called flipping (see figure 2.8 for the output).

`OpenCV` flips the image with the function `cv2.flip`.

```python
import numpy as np
import cv2 as ocv

image = ocv.imread("image.jpg")
ocv.imshow(" Original ", image)

def rotate ( image , theta , centre = None , scale =1.) :
    (h,w) = image.shape[:2]

    if centre is None :
        centre = ( w / 2 , h / 2)
    M = ocv.getRotationMatrix2D(centre ,theta ,scale)
    rotated = ocv.warpAffine(image, M, (w,h))
    return rotated

ocv.imshow(" Using rotate function ", rotate (image,45))
ocv.waitKey(0)
```

Listing 2.6 Application of the rotation operation.

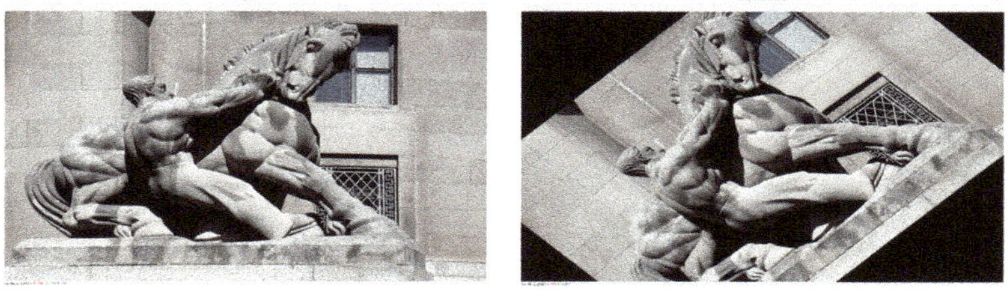

Original Image　　　　Processed Image

Figure 2.6. Output of listing 2.6: application of the rotation operation. Credit: Highsmith, Carol M., 1946 - United States Library of Congress's Prints and Photographs Division under the digital ID highsm.09920.

2.5.2 Cropping

Cropping images is very simple since images are arrays (see listing 2.9 for the code and figure 2.9 for the output).

2.5.3 Image arithmetic

Image arithmetic is similar to array arithmetic since images are arrays. However, it is vital to note that `NumPy` arrays are stored as unsigned 8 bit integers. This implies that the values of our pixels will be in the range [0, 255]. When using functions such as `cv2.add` and `cv2.subtract`, the values will be clipped to this range, even if the added or subtracted values fall outside the range of [0, 255].

2.5.4 Masking

In many applications in computer vision it is necessary to have a mask. Masking helps us to focus on specific parts of an image. A mask is an image with the same size

```
import cv2 as ocv
def resize ( image , width = None , height = None , inter = ocv.INTER_AREA ) :
    dim = None
    (h,w) = image.shape[:2]

    if width is None and height is None :
        return image

    if width is None :
        r = height / float ( h )
        dim = ( int ( w * r ) , height )

    else :
        r = width / float ( w )
        dim = ( width , int ( h * r ) )
        resized = ocv.resize ( image , dim , interpolation = inter )
        return resized

image = ocv.imread("image.jpg")
ocv.imshow(" Original ", image)
ocv.imshow("Image resized 640x480 ", resize(image,640,480))
ocv.waitKey(0)
```

Listing 2.7 Resizing operation.

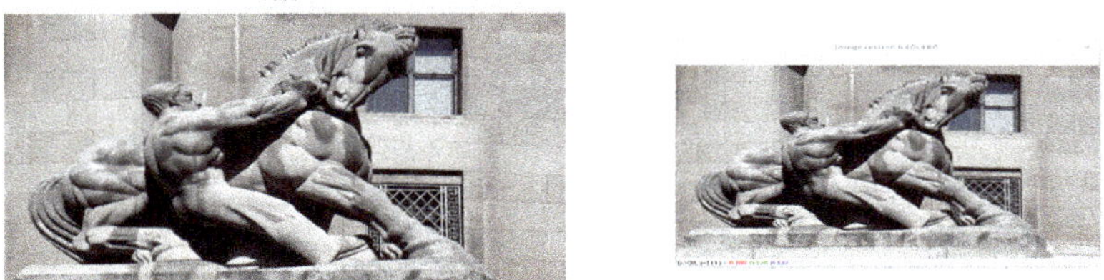

Processed Image

Original Image

Figure 2.7. Output of listing 2.7: resizing operation. Credit: Highsmith, Carol M., 1946 - United States Library of Congress's Prints and Photographs Division under the digital ID highsm.09920.

as the image that will be masked. A mask has only two-pixel values, 0 and 255 (see listing 2.10 for the code and figure 2.10 for the output).

We initialized the mask with an array of zeros. Then we obtain a white rectangle mask with coordinates (0, 90), (290, 450). The white colour comes from 255. Finally we compute the masked images called `masked` with the function `cv2.bitwise_and`.

2.6 Kernels and morphological operations

We have mentioned that images are arrays, and from images we can obtain subimages, which are smaller images. A subimage can be as small as one pixel

```python
import cv2 as ocv
image = ocv.imread("image.jpg") # load the image
ocv.imshow("␣Original␣", image)

flipped = ocv.flip(image,1)
ocv.imshow("␣Flipped␣Horizontally␣", flipped )

flipped = ocv.flip(image,0)
ocv.imshow("␣Flipped␣Vertically␣" , flipped )

flipped = ocv.flip (image,-1)
ocv.imshow("␣Flipped␣Horizontally␣&␣Vertically␣" , flipped )
ocv.waitKey(0)
```

Listing 2.8 Flipping operation.

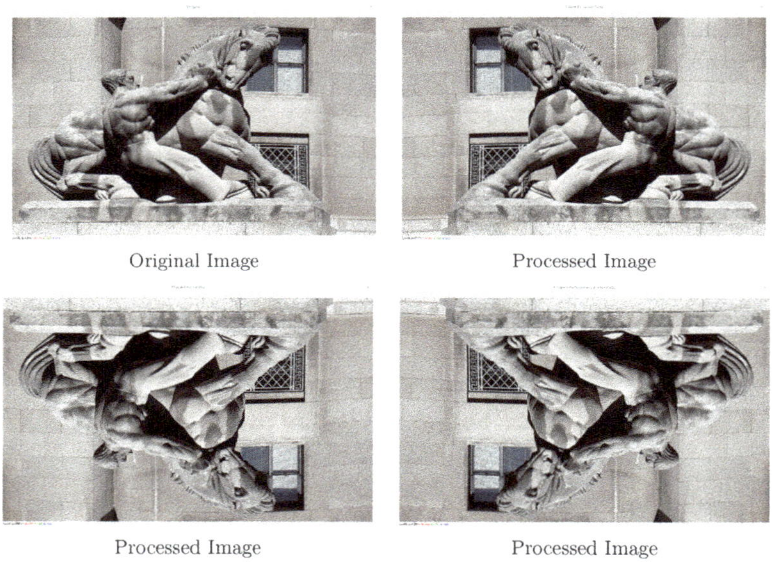

Figure 2.8. Output of listing 2.8: flipping operation. Credit: Highsmith, Carol M., 1946 - United States Library of Congress's Prints and Photographs Division under the digital ID highsm.09920.

and as large as the original image. A kernel is a subimage with specific values to perform a particular task over the image. These tasks are performed by convoluting the kernel with an image.

The convolution is the sum of the multiplications between the elements of the neighbourhood of a pixel and the elements of the kernel. When the convolution is computed over an image, it means that we apply the convolution to all the pixels of the image.

```python
import cv2 as ocv

image = ocv.imread("image.jpg")
ocv.imshow("Original", image)

crop_image = image[200:350, 300:600]

ocv.imshow("Cropped Image", crop_image)
ocv.waitKey(0)
```

Listing 2.9 Crop operation.

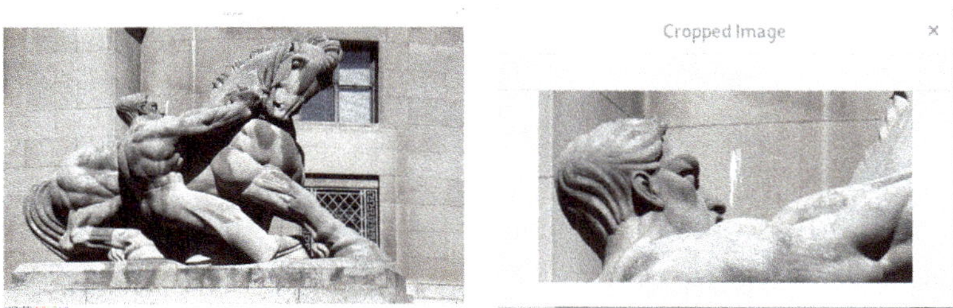

Original Image Processed Image

Figure 2.9. Output of listing 2.9: crop operation. Credit: Highsmith, Carol M., 1946 - United States Library of Congress's Prints and Photographs Division under the digital ID highsm.09920.

```python
import numpy as np
import argparse
import cv2 as ocv

image = ocv.imread("image.jpg" )
ocv.imshow(" Original ", image )

mask = np.zeros(image.shape[:2] , dtype = "uint8" )
ocv.rectangle ( mask , (600 ,20) , (960 ,450) , 255 , -1)
ocv.imshow( " Mask " , mask )

masked = ocv.bitwise_and(image , image , mask = mask )
ocv.imshow( " Mask Applied to Image " , masked )
ocv.waitKey(0)
```

Listing 2.10 Masking operation.

For example, the convolution between K and I_s is given by $K*I_s$, where

$$K = \begin{bmatrix} 1 & 0 & 2 \\ 0 & 1 & 2 \\ 0 & 1 & 0 \end{bmatrix}, \tag{2.3}$$

K is the kernel, and

Original Image Processed Image

Figure 2.10. Output of listing 2.10: masking operation. Credit: Highsmith, Carol M., 1946 - United States Library of Congress's Prints and Photographs Division under the digital ID highsm.09920.

$$I_s = \begin{bmatrix} 1 & 2 & 2 \\ 3 & 5 & 3 \\ 2 & 3 & 1 \end{bmatrix}, \tag{2.4}$$

I_s is the neighbourhood of some image. Notice that K and I_s have the same size.

First we compute the one-to-one multiplication

$$\begin{bmatrix} 1 & 0 & 2 \\ 0 & 1 & 2 \\ 0 & 1 & 0 \end{bmatrix} .* \begin{bmatrix} 1 & 2 & 2 \\ 3 & 5 & 3 \\ 2 & 3 & 1 \end{bmatrix} = \begin{bmatrix} 1 & 0 & 4 \\ 0 & 5 & 6 \\ 0 & 3 & 0 \end{bmatrix}, \tag{2.5}$$

where .* means one-to-one multiplication in our notation. Then we sum and we obtain 18. The convolution over an image is applying the same process over all the pixels of the image.

With the proper kernels and convolution, we can perform morphological operations on images. Morphological operations are simple transformations applied to binary or greyscale images. More specifically, we apply morphological operations to the shapes and structures inside images.

2.6.1 Erosion and dilatation

Erosion shrinks the white pixels in a binary image (Haralick *et al* 1987). At the same time, dilatation expands the white pixels of the binary image. The following code applies a series of erosions over an image. `grey` is the greyscale image of `image`. `image` is not in RGB but in BGR. The red and the blue are flipped. Thus we use `cv2.COLOR_BGR2GRAY` as an input for `cv2.cvtColor`. `cv2.erode` erodes `grey` three times, since `iterations=3` (see listing 2.11 for the code and figure 2.11 for the output).

The code in listing 2.12 is the same, but we change function `cv2.erode` to `cv2.dilate`, thus `grey` is dilated three times (see figure 2.12 for the output).

There are also two more basic morphological operations. A combination of erosion and dilatation generates these morphological operations. The first is opening. An opening is an erosion followed by a dilation. The second operation is closing. Closing is the opposite of opening. Thus, a closing is a dilation followed by an erosion.

```
import cv2 as ocv

image = ocv.imread ("image.jpg" )
ocv.imshow("␣Original␣",image)

gray = ocv.cvtColor(image, ocv.COLOR_BGR2GRAY )
ocv.imshow ( "␣Original␣" , image )
ocv.imshow ( "␣Gray␣" , gray )

eroded = ocv.erode(gray.copy() , None , iterations =3)
ocv.imshow( "␣Eroded␣image␣" , eroded )
ocv.waitKey(0)
```

Listing 2.11 Erosion operation.

Original Image   Processed Image

Figure 2.11. Output of listing 2.11: erosion operation. Credit: Highsmith, Carol M., 1946 - United States Library of Congress's Prints and Photographs Division under the digital ID highsm.09920.

```
import cv2 as ocv

image = ocv.imread("image.jpg" )
gray =ocv.cvtColor(image,ocv.COLOR_BGR2GRAY )
ocv.imshow("␣Gray␣", gray )

eroded = ocv.dilate(gray.copy () , None , iterations =3)
ocv.imshow( "␣Dilated␣image␣" , eroded )
ocv.waitKey(0)
```

Listing 2.12 Dilation operation.

2.7 Blurring

Smoothing and blurring are among the key pre-processing steps in all artificial vision. Blurring involves applying a type of filter over an image. It is achieved through the convolution of the kernel with an image (Shen *et al* 2007).

The kernel that has all its elements equal to $1/n$, where n is the number of elements in the kernels, is called the averaging kernel (see listing 2.13 for the code and figure 2.13 for the output).

Figure 2.12. Output of listing 2.12: dilation operation. Credit: Highsmith, Carol M., 1946 - United States Library of Congress's Prints and Photographs Division under the digital ID highsm.09920.

```
import cv2 as ocv

image = ocv.imread("image.jpg")
ocv.imshow(" Original " , image )

# convolutions with an averaging kernels of sides Ksizes
Ksizes = [(3 , 3) , (9 , 9) , (15 , 15) ]
for ( kX , kY ) in Ksizes :
    blurred = ocv.blur ( image , ( kX , kY ) )

ocv.imshow(" Averaged image with kernel ({} , {}) ".format(
   kX,kY),blurred)
ocv.waitKey (0)
```

Listing 2.13 Averaging operation.

Figure 2.13. Output of listing 2.13: averaging operation. Credit: Highsmith, Carol M., 1946 - United States Library of Congress's Prints and Photographs Division under the digital ID highsm.09920.

The `for` loop computes and displays a blurred image with kernels of size `Ksizes`. For this, the function `cv2.blur` is applied over `cv2.image`.

`OpenCV` has other kernels for blurring, such as kernels with values that have a Gaussian distribution or take the median value of the neighbourhood. The code in listing 2.14 uses a kernel with a Gaussian distribution (see figure 2.14 for the output).

```
import cv2 as ocv
image = ocv.imread ( "image.jpg" )
ocv.imshow ("␣Original␣" , image )

Ksizes = [(3 , 3) , (9 , 9) , (15 , 15) ]
for ( kX , kY ) in Ksizes :
    blurred = ocv.GaussianBlur(image,(kX,kY),0)

ocv.imshow("␣Gaussian␣image␣with␣kernel␣({}␣,␣{})␣".format(
   kX,kY),blurred)
ocv.waitKey (0)
```

Listing 2.14 Gaussian operation.

Figure 2.14. Output of listing 2.14: Gaussian operation. Credit: Highsmith, Carol M., 1946 - United States Library of Congress's Prints and Photographs Division under the digital ID highsm.09920.

```
import cv2 as ocv
image = ocv.imread ("image.jpg")
ocv.imshow( "␣Original␣" , image )

for k in (3 , 9 , 15) :
    blurred = ocv.medianBlur(image,k)

ocv.imshow("␣Median␣image␣{}␣".format(k),blurred)
ocv.waitKey (0)
```

Listing 2.15 Median operation.

Finally, we have a code (listing 2.15) that computes a blurred image with a median kernel (see figure 2.15 for the output).

2.8 Thresholding

Thresholding is the binarization of an image. This means that after the thresholding, the pixels of the image will be white or black. White if their value is 255 or black if their value is 0. Let T be our threshold. If a pixel value is greater than T, we set it to 255. Otherwise, we set it to 0. This is the case for standard binarization. In the case of inverse

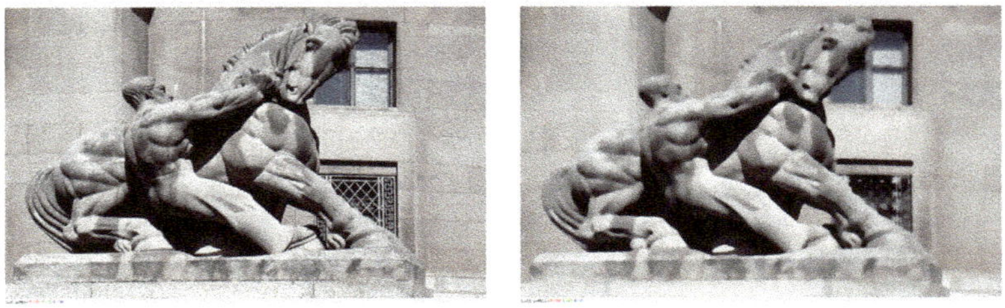

Figure 2.15. Output of listing 2.15: median operation. Credit: Highsmith, Carol M., 1946 - United States Library of Congress's Prints and Photographs Division under the digital ID highsm.09920.

```
import cv2 as ocv# load opencv

# load the image and show it
image = ocv.imread("image.jpg")
ocv.imshow("␣Original␣" , image )

# threshold
th =150;

# standard binarization
(T , thresh ) = ocv.threshold(image,th,255,ocv.
   THRESH_BINARY )
ocv.imshow("␣Standard␣binarization␣" , thresh )

# inverse binarization
(T , threshInv ) = ocv.threshold(image,th,255,ocv.
   THRESH_BINARY_INV )
ocv.imshow("␣Inverse␣binarization␣" , threshInv )
ocv.waitKey(0)
```

Listing 2.16 Threshold operation.

binarization, if the pixel value is greater than T we set it to 0, otherwise we set it to 255 (see listing 2.16 for the code and figure 2.16 for the output) (He *et al* 2013, Kapur 2017).

The code in listing 2.16 computes a standard binarization and then inverse binarization with the function `cv2.threshold`. The inputs of the function `cv2.threshold` can be `cv2.THRESH_BINARY` and `cv2.THRESH_BINARY_INV`, depending on which kind of binarization is required. `T` is the threshold, in this case, `T=150`. `thresh` is the image with standard binarization and `threshInv` is the image with inverse binarization.

There is a commonly used algorithm to compute binarizations called Otsu's method. The algorithm exhaustively searches for the threshold that minimizes the intra-class variance. Hence, for a given image Otsu's method computes a threshold for us. The code in listing 2.17 uses the input `cv2.THRESH_OTSU` to compute the threshold and the binarization (see figure 2.17 for the output) (Klette 1980).

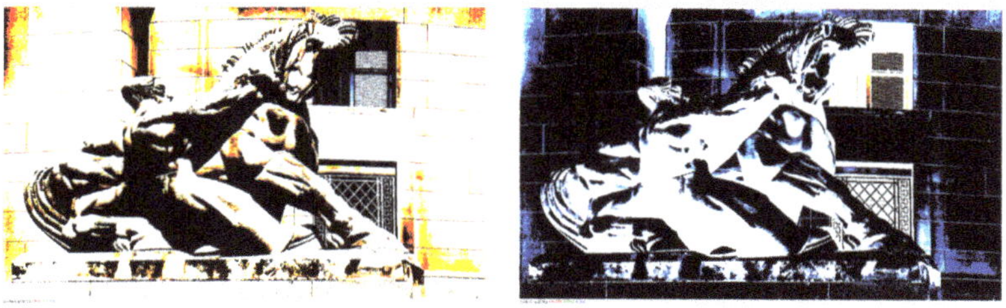

Figure 2.16. Output for listing 2.16: threshold operation; standard binarization on the left and inverse binarization on the right. Credit: Highsmith, Carol M., 1946 - United States Library of Congress's Prints and Photographs Division under the digital ID highsm.09920.

```
import cv2 as ocv # load opencv

# load the image and show it
image = ocv.imread ( "image.jpg" )

gray = ocv.cvtColor(image, ocv.COLOR_BGR2GRAY)
ocv.imshow(" Original gray ", gray)

# Otsu's binarization
(T , thresh ) = ocv.threshold(gray, 0, 255, ocv.THRESH_BINARY)
ocv.imshow(" Standar binarization " , thresh )

(T , thresh ) = ocv.threshold(gray, 0, 255, ocv.THRESH_OTSU)
ocv.imshow(" Otsu's binarization " , thresh )
ocv.waitKey(0)
```

Listing 2.17 Otsu operation.

2.9 Gradients and edge detection

2.9.1 Gradients

The Sobel operator is used in image processing, in particular in edge detection algorithms. Technically it is a discrete differential operator that calculates an approximation to the gradient of the intensity function of an image. For each pixel of the image to be processed, the result of the Sobel operator is both the corresponding gradient vector and the norm of this vector (Gao *et al* 2010, Umesh 2012).

The Sobel operator calculates the intensity gradient of an image at each pixel. Thus for each pixel this operator gives the magnitude of the greatest possible change, its direction and the direction from dark to light. The result shows how abruptly or smoothly an image changes at each analysed pixel and, consequently, how likely it is that it represents an edge in the image and also the orientation toward which that edge tends. In practice, the calculation of the magnitude—the probability of an

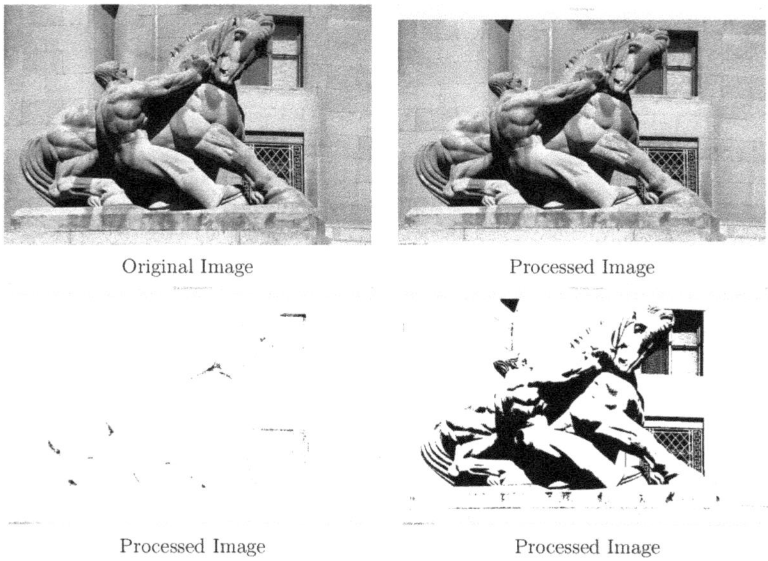

Figure 2.17. Output for listing 2.17: Otsu operation. Credit: Highsmith, Carol M., 1946 - United States Library of Congress's Prints and Photographs Division under the digital ID highsm.09920.

edge—is more reliable and easier to interpret than the calculation of the direction from dark to white (Vincent *et al* 2009).

Mathematically, the gradient of a function of two variables (in this case, the intensity function of the image) for each pixel is a two-dimensional vector whose components are given by the first derivatives of the vertical and horizontal directions. For each pixel in the image the gradient vector points in the direction of the maximum possible increase in intensity and the magnitude of the gradient vector corresponds to the amount of change in intensity in that direction.

What has been said in the previous paragraphs implies that the result of applying the Sobel operator on a region with constant image intensity is a zero vector, and the result of applying it at a pixel on an edge is a vector that crosses the edge (perpendicular) whose meaning is from the darkest points to the lightest. Mathematically, the operator uses two 3×3 element kernels to apply a convolution to the original image to calculate approximations for the derivatives, one kernel for the horizontal changes and another for the vertical ones.

The derivative of image I in x is

$$G_x = I^* \begin{bmatrix} -1 & 0 & +1 \\ -2 & 0 & +2 \\ -1 & 0 & +1 \end{bmatrix}. \tag{2.6}$$

The derivative of image I in y is

$$G_y = I^* \begin{bmatrix} -1 & -2 & -1 \\ 0 & 0 & 0 \\ +1 & +2 & +1 \end{bmatrix}. \tag{2.7}$$

Thus the gradient is given by

$$G = \sqrt{G_x + G_y}. \tag{2.8}$$

With this information we can also calculate the direction of the gradient

$$\Theta = \arctan\left(\frac{G_x}{G_y}\right), \tag{2.9}$$

where, for example, Θ is 0 for vertical edges with darker points on the left side.

The code in listing 2.18 computes the derivatives of the image with Sobel operators (see figure 2.18 for the output).

The function `cv2.Sobel` computes the derivative in the x or y direction, depending the requirements of the user. With `cv2.convertScaleAbs` we convert the floating point data type to an unsigned 8 bit. Thus other `OpenCV` functions can use the images `gx` and `gy`. Finally we compute the gradient with `cv2.addWeighted`.

```
import cv2 as ocv # load opencv

# load the image and show it
image = ocv.imread("image.jpg")
ocv.imshow(" Original ",image)

gray =ocv.cvtColor(image,ocv.COLOR_BGR2GRAY)
ocv.imshow( " Gray " , gray )

# compute derivative along the X
gX = ocv.Sobel( gray ,ddepth = ocv.CV_64F , dx =1 , dy =0)
# compute derivative along the Y
gY = ocv.Sobel( gray ,ddepth = ocv.CV_64F , dx =0 , dy =1)
# convert the floating point data type to an unsigned 8 -
    bit

gX = ocv.convertScaleAbs(gX)
gY = ocv.convertScaleAbs(gY)

# compute the gradient with the
gradient = ocv.addWeighted(gX,0.5,gY,0.5,0)

# show our output images
ocv.imshow ( " Sobel X " , gX )
ocv.imshow ( " Sobel Y " , gY )
ocv.imshow ( " Gradient " , gradient )
ocv.waitKey (0)
```

Listing 2.18 Sobel operation.

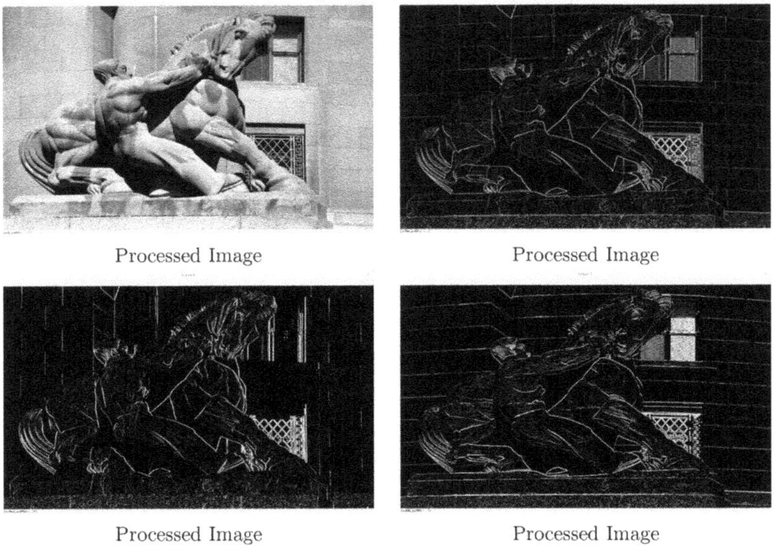

Figure 2.18. Output for listing 2.18: Sobel operation. Credit: Highsmith, Carol M., 1946 - United States Library of Congress's Prints and Photographs Division under the digital ID highsm.09920.

2.9.2 Edges

Canny's algorithm is an operator developed by John F Canny in 1986 that uses a multi-stage algorithm to detect a wide range of edges in images. Most importantly, Canny also developed a computational theory about edge detection, which explains why the technique works. Canny's purpose was to discover the optimal edge detection algorithm. For an edge detector to be considered optimal, it must meet the following criteria:

1. *Good detection*: the algorithm should mark as many real numbers on the edges of the image as possible.
2. *Good localization*: the marking edges should be as close to the edge of the actual image as possible.
3. *Minimum response*: the edge of an image should only be marked once and, whenever possible, image noise should not create false boundaries.

To meet these requirements, Canny uses variance calculus—a function-finding technique that optimizes a given functional.

The code in listing 2.19 is an example of the implementation of Canny's algorithm using the function `cv2.Canny`. Note that `c1`, `c2`, `c3` and `c4` are the resulting images of Canny's algorithm using different thresholds (see figure 2.19 for the output).

The stages of Canny's algorithm are as follows:

1. *Noise reduction*: Canny's edge detection algorithm uses a filter based on the first derivative of a Gaussian. Since it is susceptible to noise present in raw image data, the original image is transformed with a Gaussian filter.

```
import cv2 as ocv

image = ocv.imread ( "image.jpg" )
gray = ocv.cvtColor ( image ,ocv.COLOR_BGR2GRAY )
blurred = ocv.GaussianBlur ( gray , (5 , 5) , 0)

ocv.imshow(" Original " , image )
ocv.imshow(" Blurred " , blurred )

# compute edges for different thresholds
c1 = ocv.Canny(blurred , 15 , 200)
c2 = ocv.Canny(blurred , 70 , 150)
c3 = ocv.Canny(blurred , 200 , 250)
c4 = ocv.Canny(blurred , 210 , 250)

ocv.imshow("c1", c1 )
ocv.imshow("c2", c2 )
ocv.imshow("c3", c3 )
ocv.imshow("c4", c4 )
ocv.waitKey(0)
```

Listing 2.19 The Canny operation.

The result is a slightly blurred image compared to the original version. This new image is not affected by a single pixel of noise to a significant degree.

2. *Finding the intensity of the image gradient*: The edge of an image can point in different directions, so Canny's algorithm uses four filters to detect the horizontal, vertical and diagonal at the edges of the blurred image. The edge detection operator (Sobel) returns a value for the first derivative in the horizontal direction G_y and the vertical direction G_x. From this, the edge gradient and direction can be determined.

2.10 Histograms

A histogram is a graphical representation of a variable in the form of bars, where the area of each bar is proportional to the frequency of the represented values. In the histogram of an image the represented values are the values of the pixels. Recall that the values of the pixels go from 0 to 255. Thus a histogram of an image counts how many pixels are of a certain value in the image (Xu *et al* 2011).

Histograms are used to obtain a general panorama of the distribution of the values of the pixels. In this way it offers an overview, allowing us to observe a preference, or trend, on the part of the subimage or image. Thus we can show the behaviour of a colour in an image. We can observe the degree of homogeneity in the image. We can also see the agreement between the values of the pixels of all the parts that make up the image or a subimage. Finally, in contrast, we are able to observe the degree of variability and, therefore, the dispersion of all the values.

The code in listing 2.20 shows a histogram of a greyscale image in `Python` (see figure 2.20 for the output).

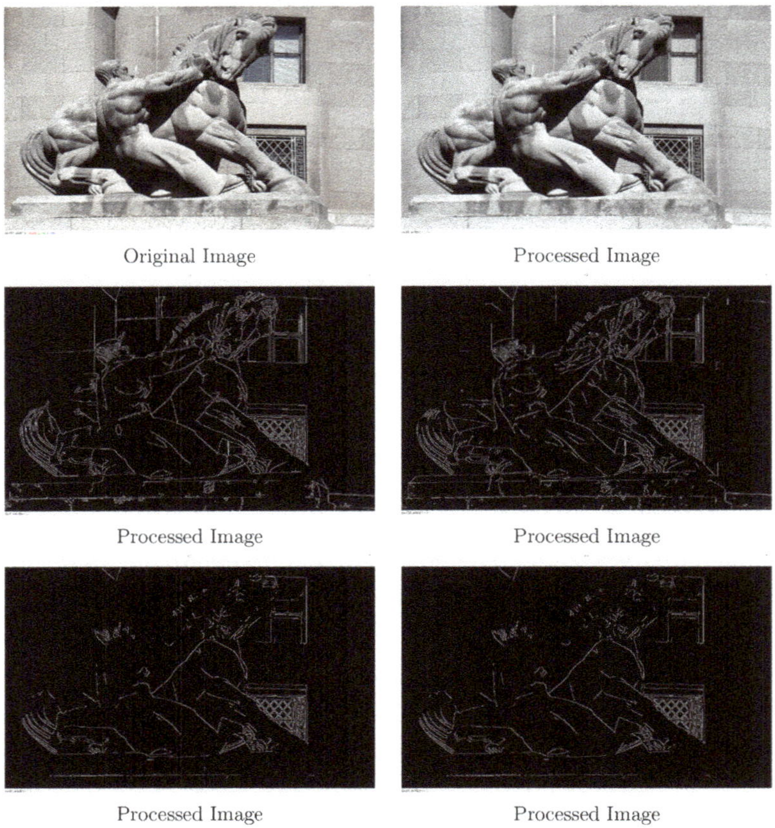

Figure 2.19. Output for listing 2.19: the Canny operation. Credit: Highsmith, Carol M., 1946 - United States Library of Congress's Prints and Photographs Division under the digital ID highsm.09920.

```
import cv2 as ocv
from matplotlib import pyplot as plot

image = ocv.imread("image.jpg")
ocv.imshow(" Original ", image)

gray= ocv.cvtColor(image,ocv.COLOR_BGR2GRAY)
ocv.imshow("Grayscale", gray)
ocv.waitKey(0)

# get the grayscale histogram of image
hist = ocv.calcHist([gray], [0] , None , [256] , [0 , 256])

plot.figure()
plot.title( " Grayscale Histogram " )
plot.xlabel( " Bins " )
plot.ylabel( " # of Pixels " )
plot.xlim([0 , 256])
plot.plot( hist )
plot.show()
```

Listing 2.20 Histogram operation.

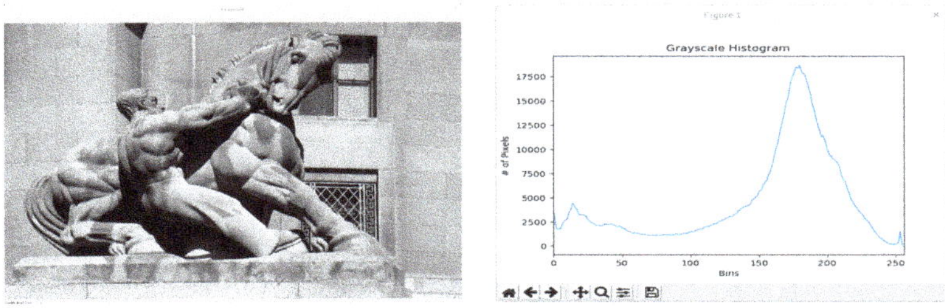

Processed Image　　　　Processed Image

Figure 2.20. Output for listing 2.20: histogram operation. Credit: Highsmith, Carol M., 1946 - United States Library of Congress's Prints and Photographs Division under the digital ID highsm.09920.

```
import cv2 as ocv
from matplotlib import pyplot as plot # load pyplot


image = ocv.imread( "image.jpg" )
ocv.imshow(" Original ", image)
ocv.waitKey(0)

channels = ocv.split(image)
colors = ( "b" , "g" , "r" )

plot.figure ()
plot.title ( " Color Histogram " )
plot.xlabel ( " Bins " )
plot.ylabel ( " # of Pixels " )


for(channel,color) in zip ( channels , colors ) :
    hist =ocv.calcHist([channel] , [0] , None , [256] , [0 , 256])
    plot.plot(hist , color = color )
    plot.xlim([0 , 256])
    plot.show()
```

Listing 2.21 Colour histogram operation.

`cv2.calcHist` computes the histogram of `image`. Since `image` is greyscale it has only one channel and this 0 is an input of `cv2.calcHist`. We want 256 bins in the histogram, so we use the range from 0 to 255.

The code in listing 2.21 presents a histogram of a colour image (see figure 2.21 for the output).

To compute the histogram of `image`, we first need to separate its BRG channels using `split`. Then we compute each histogram with the `for` loop.

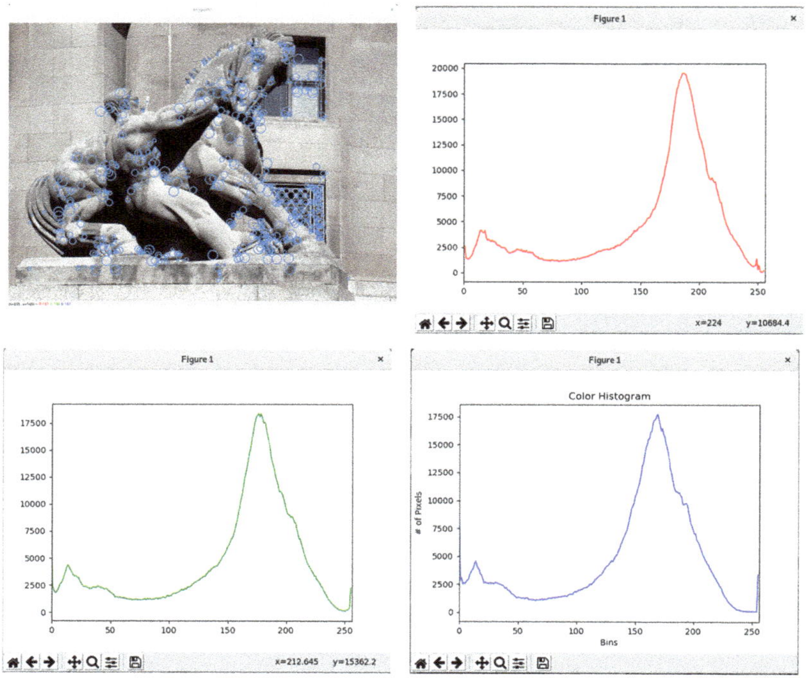

Figure 2.21. Output for listing 2.21: colour histogram operation. Credit: Highsmith, Carol M., 1946 - United States Library of Congress's Prints and Photographs Division under the digital ID highsm.09920.

2.11 End notes

In this chapter we studied the basic properties and manipulation of images. We discussed several algorithms and their respective codes in `Python`. Examples of these algorithms are translations, rotations, cropping, thresholding, gradients, edge detectors, histograms, contours, etc.

References

Chityala R and Pudipeddi S 2020 *Image Processing and Acquisition Using* `Python` (Boca Raton, FL: CRC Press)

Forsyth D A and Ponce J 2002 *Computer Vision: A Modern Approach* (Englewood Cliffs, NJ: Prentice Hall)

Gao W, Zhang X, Yang L and Liu H 2010 An improved Sobel edge detection *2010 3rd Intr. Conf. on Computer Science and Information Technology* (Piscataway, NJ: IEEE) pp 67–71

Gonzalez R C, Woods R E and Eddins S L 2004 *Digital Image Processing Using MATLAB* (New Delhi: Pearson Education India)

Granlund G H and Knutsson H 2013 *Signal Processing for Computer Vision* (Berlin: Springer)

Haralick R M, Sternberg S R and Zhuang X 1987 Image analysis using mathematical morphology *IEEE Trans. Pattern Anal. Mach. Intell.* **PAMI-9** 532–50

He Z-Y, Sun L-N and Chen L-G 2013 Fast computation of threshold based on Otsu criterion *Dianzi Xuebao (Acta Electron. Sin.)* **41** 267–72

Kapur S 2017 *Computer Vision with* `Python` *3* (Birmingham: Packt)

Klette R 1980 Parallel operations on binary images *Comput. Graph. Image Process.* **14** 145–58

Klette R 2014 *Concise Computer Vision* (Berlin: Springer)

López A F J, Pelayo M C P and Forero Á R 2016 Teaching image processing in engineering using `Python` *IEEE Rev. Iberoam. Tecnol. Aprendizaje.* **11** 129–36

Lutz M 2001 *Programming* `Python` (Sebastopol, CA: O'Reilly)

Oliphant T E 2007 `Python` for scientific computing *Comput. Sci. Eng.* **9** 10–20

Raschka S 2015 `Python` *Machine Learning* (Birmingham: Packt)

Schalkoff R J 1989 *Digital Image Processing and Computer Vision* vol 286 (New York: Wiley)

Shen B, Xu Y, Lu G and Zhang D 2007 Detecting iris lacunae based on Gaussian filter *Third Intr. Conf. on Intelligent Information Hiding and Multimedia Signal Processing (IIH-MSP 2007)* vol 1 (Piscataway, NJ: IEEE) pp 233–236

Szeliski R 2010 *Computer Vision: Algorithms and Applications* (Berlin: Springer)

Umesh P 2012 Image processing in `Python` *CSI Communications* **23** 2

Van Rossum G and Drake F L Jr 1995 *Python Tutorial* vol 620 (Amsterdam: Centrum voor Wiskunde en Informatica)

Vincent O R *et al* 2009 A descriptive algorithm for Sobel image edge detection *Proc. of Informing Science and IT Education Conf. (InSITE)* vol 40 (Santa Rosa, CA: Informing Science Institute) pp 97–107

Xu X, Xu S, Jin L and Song E 2011 Characteristic analysis of Otsu threshold and its applications *Pattern Recognit. Lett.* **32** 956–61

IOP Publishing

Optics and Artificial Vision

Rafael G González-Acuña, Héctor A Chaparro-Romo and Israel Melendez-Montoya

Chapter 3

Optical flow

In this chapter we study the main optical flow algorithms, the Lucas–Kanade algorithm and the Horn–Schunck algorithm. We deduce them step by step, then we show their main properties and finally we present examples of them in `Python`.

3.1 Introduction

Optical flow is the pattern of the apparent movement of objects in the image between two consecutive frames caused by the movement of the object or camera. It is a 2D vector field where each vector is a displacement vector showing the direction of the points from the first frame to the second. Therefore, optical flow studies the vector displacement between frames or images. Usually these images are in greyscale. The optical flow works on two main assumptions: the pixel intensities of an object do not change between consecutive frames and neighbouring pixels have similar motion (Bauer *et al* 2006, Beauchemin and Barron 1995, Fleet and Weiss 2006, Grant 1997).

In this chapter we are going to explore some of the most popular optical flow algorithms, the Lucas–Kanade algorithm and the Horn–Schunck algorithm. Both compute the vertical and horizontal displacement of the pixels from two consecutive images. The Lucas–Kanade algorithm works on a local approach, while the Horn–Schunck algorithm uses a global approach to the image.

3.2 The Lucas–Kanade algorithm

The Lucas–Kanade algorithm has the aim to calculate the displacement of particles using two consecutive images. The first image $I_1(x, y)$ is at time t, while the second image $I_2(x, y)$ is at time $t + dt$, where dt is the time between the images.

The Lucas–Kanade algorithm inputs the two images $I_1(x, y)$ and $I_2(x, y)$ into a system of differential equations to find the displacement components u and v. u is the horizontal displacement and v is the vertical displacement (Bouguet *et al* 2001, Bruhn *et al* 2005).

doi:10.1088/978-0-7503-3707-6ch3

3.2.1 Assumptions

Before we start with the step by step derivation of the Lucas–Kanade algorithm, it is useful to note some assumptions implemented in the algorithm. As mentioned before, the goal of the Lucas–Kanade algorithm is to find the components of displacement of u and v. The code makes the following assumptions: the first image I_1 is similar to the second image I_2; u and v are relatively small; and $I_1(x + u, y + v)$ must be similar to $I_2(x, y)$. The objective is to find u and v such that the difference between $I_1(x + u, y + v)$ and $I_2(x, y)$ is a minimum. This means that the change over time from the first image I_1 to the second image I_2 is smooth without discontinuities (Baker and Matthews 2004, Baker *et al* 2003).

3.2.2 The theory behind the Lucas–Kanade algorithm

The Lucas–Kanade algorithm is based on the least-squares criterion. The least-squares criterion is a standard approach in regression analysis to approximate the solution of overdetermined systems by minimizing the sum of the squares of the residuals produced in the results of every single equation. Remember that an overdetermined system is a set of equations in which there are more equations than unknowns. Hence, the Lucas–Kanade algorithm computes the squared error between the images and obtains displacement components u and v such that the error is minimized (Rav-Acha and Peleg 2006, Schreiber 2008, Tong 2013).

Equation (3.1) is the squared error E over a neighbourhood of size $N \times M$ pixels, where $I_1(x + u, y + v)$ is the first image evaluated at the displacement components u and v and $I_2(x, y)$ is the second image:

$$E^2 = \sum_{i=1}^{N}\sum_{j=1}^{M}[I_1(x + i + u, y + j + v) - I_2(x + i, y + j)]^2. \tag{3.1}$$

To minimize the above equation it is better to express $I_1(x + u, y + v)$ in a Taylor series. Thus we take the first three terms of the Taylor series of $I_1(x + u, y + v)$, note that we obtain the linear terms

$$I_1(x + u, y + v) = I_1(x, y) + u\frac{\partial I_1(x, y)}{\partial x} + v\frac{\partial I_1(x, y)}{\partial y}. \tag{3.2}$$

Replacing equation (3.2) in equation (3.1) we obtain

$$\begin{aligned} E^2 = \sum_{i=1}^{N}\sum_{j=1}^{M}\Bigg[& I_1(x + i, y + j) + u\frac{\partial I_1(x + i, y + j)}{\partial x} \\ & + v\frac{\partial I_1(x + i, y + j)}{\partial y} - I_2(x + i, y + j)\Bigg]^2. \end{aligned} \tag{3.3}$$

To minimize the function E^2 we apply a derivative with respect to u and v and equalize to zero. We start with the derivative with respect to u:

$$\frac{\partial E}{\partial u} = 2\sum_{i=1}^{N}\sum_{j=1}^{M}\left[I_1(x+i,\,y+j) + u\frac{\partial I_1(x+i,\,y+j)}{\partial x} + v\frac{\partial I_1(x+i,\,y+j)}{\partial y}\right. \\ \left. - I_2(x+i,\,y+j)\right] \times \frac{\partial I_1(x+i,\,y+j)}{\partial x} = 0. \tag{3.4}$$

Then we follow with the derivative with respect to v:

$$\frac{\partial E}{\partial v} = 2\sum_{i=1}^{N}\sum_{j=1}^{M}\left[I_1(x+i,\,y+j) + u\frac{\partial I_1(x+i,\,y+j)}{\partial x} + v\frac{\partial I_1(x+i,\,y+j)}{\partial y}\right. \\ \left. - I_2(x+i,\,y+j)\right] \times \frac{\partial I_1(x+i,\,y+j)}{\partial y} = 0. \tag{3.5}$$

Now we need to expand equations (3.4) and (3.5). We shall begin with equation (3.4):

$$0 = \sum_{i=1}^{N}\sum_{j=1}^{M} I_1(x+i,\,y+j)\frac{\partial I_1(x+i,\,y+j)}{\partial x} + u\sum_{i=1}^{N}\sum_{j=1}^{M}\left[\frac{\partial I_1(x+i,\,y+j)}{\partial x}\right]^2 \\ + v\sum_{i=1}^{N}\sum_{j=1}^{M}\frac{\partial I_1(x+i,\,y+j)}{\partial x}\frac{\partial I_1(x+i,\,y+j)}{\partial y} - \sum_{i=1}^{N}\sum_{j=1}^{M} I_2(x+i,\,y+j)\frac{\partial I_1(x+i,\,y+j)}{\partial x}. \tag{3.6}$$

Then we expand equation (3.5):

$$0 = \sum_{i=1}^{N}\sum_{j=1}^{M} I_1(x+i,\,y+j)\frac{\partial I_1(x+i,\,y+j)}{\partial y} + v\sum_{i=1}^{N}\sum_{j=1}^{M}\left[\frac{\partial I_1(x+i,\,y+j)}{\partial y}\right]^2 \\ + u\sum_{i=1}^{N}\sum_{j=1}^{M}\frac{\partial I_1(x+i,\,y+j)}{\partial x}\frac{\partial I_1(x+i,\,y+j)}{\partial y} - \sum_{i=1}^{N}\sum_{j=1}^{M} I_2(x+i,\,y+j)\frac{\partial I_1(x+i,\,y+j)}{\partial y}. \tag{3.7}$$

Now note that from equations (3.6) and (3.7) we can obtain a differential-algebraic equation system that can be expressed in matrix form. We may call it a differential-algebraic equation system, but it is in fact not one. Because the unknowns are u and v, the derivatives of the images can be computed with the algorithms presented in chapter 2. Using the proper kernel we can obtain $\partial I_1/\partial x$ and $\partial I_1/\partial y$.

Returning to equations (3.6) and (3.7), they can be represented in the following matrix form:

$$\begin{bmatrix} \sum_{i=1}^{N}\sum_{j=1}^{M}\left(\frac{\partial I_1}{\partial x}\right)^2 & \sum_{i=1}^{N}\sum_{j=1}^{M}\frac{\partial I_1}{\partial x}\frac{\partial I_1}{\partial y} \\ \sum_{i=1}^{N}\sum_{j=1}^{M}\frac{\partial I_1}{\partial x}\frac{\partial I_1}{\partial y} & \sum_{i=1}^{N}\sum_{j=1}^{M}\left(\frac{\partial I_1}{\partial y}\right)^2 \end{bmatrix}\begin{bmatrix} u \\ v \end{bmatrix} = \begin{bmatrix} \sum_{i=1}^{N}\sum_{j=1}^{M} I_2\frac{\partial I_1}{\partial x} - \sum_{i=1}^{N}\sum_{j=1}^{M} I_1\frac{\partial I_1}{\partial x} \\ \sum_{i=1}^{N}\sum_{j=1}^{M} I_2\frac{\partial I_1}{\partial y} - \sum_{i=1}^{N}\sum_{j=1}^{M} I_1\frac{\partial I_1}{\partial y} \end{bmatrix}. \tag{3.8}$$

Equation (3.8) is similar to a typical system of linear equations with the form

$$Ax = b. \tag{3.9}$$

To solve equation (3.9) we multiply both sides of the equation by the inverse matrix of A. The same procedure can be used to solve equation (3.8):

$$\begin{bmatrix} u \\ v \end{bmatrix} = \begin{bmatrix} \sum_{\substack{i=1\\j=1}}^{N,M} \left(\frac{\partial I_1}{\partial x}\right)^2 & \sum_{\substack{i=1\\j=1}}^{N,M} \frac{\partial I_1}{\partial x}\frac{\partial I_1}{\partial y} \\ \sum_{\substack{i=1\\j=1}}^{N,M} \frac{\partial I_1}{\partial x}\frac{\partial I_1}{\partial y} & \sum_{\substack{i=1\\j=1}}^{N,M} \left(\frac{\partial I_1}{\partial y}\right)^2 \end{bmatrix}^{-1} \begin{bmatrix} \sum_{\substack{i=1\\j=1}}^{N,M} I_2\frac{\partial I_1}{\partial x} - \sum_{\substack{i=1\\j=1}}^{N,M} I_1\frac{\partial I_1}{\partial x} \\ \sum_{\substack{i=1\\j=1}}^{N,M} I_2\frac{\partial I_1}{\partial y} - \sum_{\substack{i=1\\j=1}}^{N,M} I_1\frac{\partial I_1}{\partial y} \end{bmatrix}. \tag{3.10}$$

The solution of the system of equations is u and v. These u and v are just for the neighbourhood of length $N \times M$. This means that for a given neighbourhood of size $N \times M$, the mean displacement values are u and v. Thus it is essential in the implementation of the algorithm to consider the size of the neighbourhood. If the neighbourhood is too large u and v will be too averaged and details may be lost. On the other hand, if the neighbourhood is too small, u and v can be a bad approximation, because the moving pixel that appears in neighbourhood of I_1 may not appear in the neighbourhood of I_2. Hence, the size of the neighbourhood is proportional to the displacements u and v.

3.2.3 The Lucas–Kanade algorithm step by step

It is important to see that equation (3.10) is an approximate solution. To obtain the best solution, the mathematical background explained in the last section must be implemented recursively. This recursive approach compensates for the use of only the linear terms in the Taylor expansion. Thus the first iteration gives a solution of the type $I_1(x + u, y + v)$; then we apply the same process considering the previous answer to find a new $I_1(x + u, y + v)$, and so on. This implies that the subimage composed of the $M \times N$ particles must be deformed according to the calculated displacement and this deformed subimage is the one used in the next iteration of the algorithm[1].

The algorithm of Lucas–Kanade for image processing.

1. Obtain the two subimages $I_1(x, y)$ and $I_2(x, y)$.
2. Take the first derivative of the first image $I_1(x, y)$ with respect to x and y.
3. Obtain the summations implied in equation (3.10).
4. Calculate u and v.
5. With the new u and v, interpolate $I_1(x, y)$ to obtain $I_1(x + u, y + v)$.
6. The $I_1(x + u, y + v)$ becomes the new $I_1(x, y)$.
7. Go to step 2 and repeat the steps until the difference in displacement for two consecutive iterations is less than a certain threshold.

[1] Note that the word subimage is the same as neighbourhood for the purposes of this book.

3.2.4 Failures of the Lucas–Kanade algorithm

There are some occasions when the Lucas–Kanade algorithm fails; the situations are listed in the following:

1. *Noisy images*. When the images are corrupted by noise or are saturated, it is difficult for the method to yield accurate results since pixel information is lost. This problem can be alleviated by preprocessing the images.
2. *Images with low movement density*. When the image $I_1(x, y)$ has almost no movement, the averaged value may be close to zero in the whole image. This problem is fixed with a proper size of the subimage.
3. *Movements with relatively large speed*. The algorithm obtains misleading results when the moving pixel is inside the neighbourhood of $I_1(x, y)$ but not inside the corresponding neighbourhood of $I_2(x, y)$. This happens when the optical flow reaches high speeds. The problem is that $I_1(x, y)$ is no longer similar to $I_2(x, y)$.

3.3 Application of the Lucas–Kanade algorithm and its `Python` code

In order to test the Lucas–Kanade algorithm in `Python`, two synthetic particle images are generated and the Lucas–Kanade algorithm is applied. The synthetic particle images are greyscale images filled with particles. They are normally used in particle image velocimetry (PIV). PIV is an optical method of flow visualization used to research fluids. It is used to obtain instantaneous velocity measurements and related properties in fluids. The fluid is seeded with tracer particles which, for sufficiently small particles, are assumed to follow the flow dynamics faithfully. To test algorithms in PIV it is common to use synthetic particle images generated by an artificial particle image generator (Raffel *et al* 2018, Santiago *et al* 1998, Westerweel 1997, Willert and Gharib 1991).

The artificial particle image generator is a code that generates images with artificial particles with known parameters: shape, diameter, dynamic range, spatial density and image bit depth. The artificial particle image generator presented here works with greyscale images and individual particles described by a Gaussian intensity profile,

$$A(x, y) = A_0\, e^{\frac{-8[(x-x_0)^2-(y-y_0)^2]}{d_r^2}}, \tag{3.11}$$

where $A(x, y)$ is the intensity at point (x, y). The centre of the particles is located at (x_0, y_0) with a peak of intensity of A_0 and d_r is the diameter of the particle.

The code (see listing 3.1) first receives the positions of the centre of every artificial particle, then with equation (3.11) it renders the intensity of every particle along with the image. As can be seen in equation (3.11) the intensity turns to zero for a position far from the centre of the particle, so the next step is to sum all the plots in one image (matrix), like a superposition of waves. The result is a single image that contains the corresponding intensity of every artificial particle.

The above description explains how the artificial particle image generator code creates the initial image I_1. Now to obtain the second image I_2 the particle image generator code moves the centre of every artificial particle by a previously known

```
# Lucas Kanade
from matplotlib import pyplot as plot
from scipy import signal
import scipy
import numpy as np
import cv2 as ocv # load opencv

def particle_generator(N_particles,X,Y,x_delta,y_delta,r):
    x = np.random.randint(0,X-1,N_particles)
    y = np.random.randint(0,X-1,N_particles)
    x2 = x + x_delta
    y2 = y + y_delta
    A =255 # amplitude

    im = np.zeros([Y,X])
    im2 = np.zeros([Y,X])

    for p in range(N_particles) :
        for y1 in range(X) :
            for x1 in range(Y) :
                im[y1,x1]=im[y1,x1]+A*np.exp
                (-2.6*((x1-x[p])**2+(y1-y[p])**2)/r**2)
                im2[y1,x1]=im2[y1,x1]+A*np.exp
                (-2.6*((x1-x2[p])**2+(y1-y2[p])**2)/r**2)
    return im , im2

# imwarp does not work
def imwarp (u ,v , im ) :
    # [ im2 ] = imWarp (u ,v , im2 ) ;
    r , c = im.shape
    x , y = np.meshgrid(np.linspace(1,r,r),
                np.linspace (1,c,c) )
    f = scipy.interpolate.interp2d(x+u,y+v,im)
    im2 = f(np.linspace(0,r,r),np.linspace (0,c,c) )
    return im2

im1,im2 = particle_generator(100,240,240,3,4,3)
ocv.imshow("Artificial Image 1", im1)
ocv.imshow("Artificial Image 2", im2)
ocv.waitKey(0)

r , c = im1.shape

# Implementing Lucas Kanade Method
ww =100;# size of the window
w = 50; # np . round ( ww /2) # size of the window
u = np.zeros([r,c]) # horizontal displecement optical flow
v = np.zeros([r,c]) # vertical displecement optical flow
uu = np.zeros([r,c])
vv = np.zeros([r,c])
```

Listing 3.1 The Lucas–Kanade operation.

```
Ix_m = signal.convolve(im1,[[-1, 1] ,[-1,1]] ,
                mode ='valid') # partial on x
Iy_m = signal.convolve(im1,[[-1,-1] ,[ 1,1]] ,
                mode ='valid') # partial on y
It_m = signal.convolve(im1, np.ones([2,2]) ,
                mode = 'valid') + \

       signal.convolve(im2,-np.ones([2,2]) ,
       mode = 'valid') # partial on t

for iter in range (10) :
    for i in range (w,r-w+1) :
        for j in range (w ,c - w +1) :
            Ix = Ix_m [i - w : i +w , j - w : j + w ]
            Iy = Iy_m [i - w : i +w , j - w : j + w ]
            It = It_m [i - w : i +w , j - w : j + w ]
            Ix = np.reshape(Ix,Ix.size)

            # we change the matriz to a vector
            Iy = np.reshape(Iy,Iy.size)

            # we change the matriz to a vector
            b = np.reshape(It,It.size)

             # we change the matriz to a vector
            A = np.column_stack((Ix,Iy))

            # A = [ Ix Iy ]; % get A here

            nu = np.dot(np.linalg.pinv(A),b)

            # nu = (( A * A ) ^ -1) * A * b ;

            u[i,j]= nu[0]
            v[i,j]= nu[1]

    uu = uu + u /( iter +1)
    vv = vv + v /( iter +1)

# Plot flow field downsize u and v

u_deci = -uu[0:r:w,0:c:w]
v_deci =  vv[0:r:w,0:c:w]

r1,c1 = v_deci.shape
# get coordinate for u and v in the original frame
X , Y = np.meshgrid(np.linspace(0,r,r+1) ,
              np.linspace(0,c,c+1))

X_deci = X[0:r:w,0:c:w]
Y_deci = Y[0:r:w,0:c:w]
```

Listing 3.1 (Continued.)

```
# draw vectors
plot.quiver(X_deci,Y_deci,u_deci,v_deci,units ='width')
plot.show()
```

Listing 3.1 (Continued.)

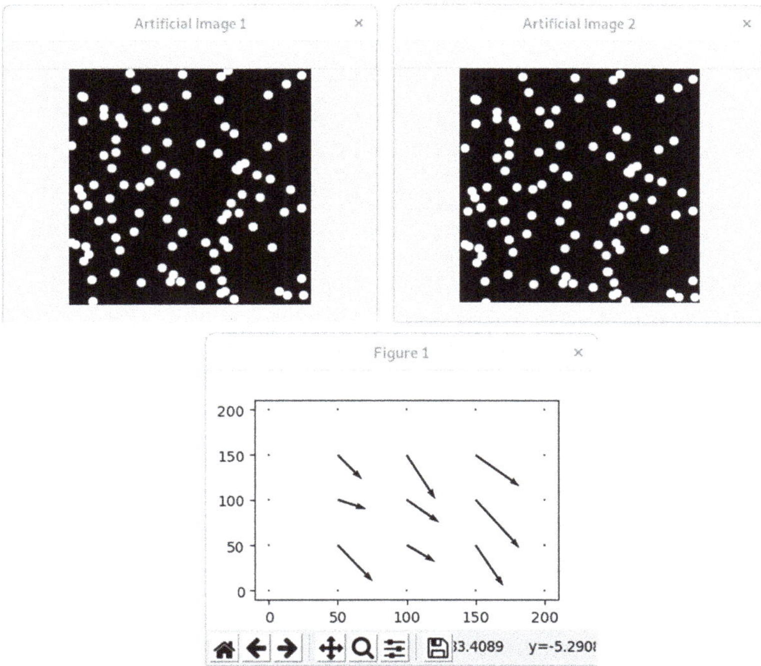

Figure 3.1. Output of listing 3.1: the Lucas–Kanade operation.

displacement. Then the code repeats the same steps that generated the first image I_1. Figure 3.1 shows typical images I_1 and I_2 that are generated by the artificial particle image generator code.

3.4 The optical flow model

There is an alternative form of the Lucas–Kanade algorithm which we study in this section. This alternative form of the Lucas–Kanade algorithm is sometimes called the optical flow model, which we are going to use here. The optical flow model has a simpler notation and it is easier to programme than the minimization approach described in the previous section. However, both are equivalent, as is shown next (Yegavian *et al* 2016).

The same assumptions as for the Lucas–Kanade algorithm are taken for the optical flow model:

1. The first image $I_1(x, y)$ is similar to the second image $I_2(x, y)$.

2. The displacements u and v are relatively small.
3. The first image $I_1(x, y)$ evaluated at $x + u$ and $y + v$ is equal to $I_2(x, y)$, $I_1(x + u, y + v) = I_2(x, y)$.

Then, by assuming an exact differential (with small displacements) for the first image,

$$I_1(x + u, y + v, t + dt) = I_1(x, y, t) + \frac{\partial I_1(x, y, t)}{\partial x}u + \frac{\partial I_1(x, y, t)}{\partial y}v + \frac{\partial I_1(x, y, t)}{\partial t}t. \quad (3.12)$$

Since the displaced image is assumed to be an exact copy of the first one, then

$$\frac{\partial I_1(x, y, t)}{\partial x}u + \frac{\partial I_1(x, y, t)}{\partial y}v + \frac{\partial I_1(x, y, t)}{\partial t}t = 0, \quad (3.13)$$

or

$$\frac{\partial I_1(x, y, t)}{\partial x}u + \frac{\partial I_1(x, y, t)}{\partial y}v = -\frac{\partial I_1(x, y, t)}{\partial t}t. \quad (3.14)$$

The derivative with respect to time is just the difference between the first and the second image. Thus we can remove the dependence of time t from both images. Taking this into consideration, from equation (3.14) we have

$$\frac{\partial I_1(x, y)}{\partial x}u + \frac{\partial I_1(x, y)}{\partial y}v = -[I_2(x, y) - I_1(x, y)]. \quad (3.15)$$

Equation (3.15) is expressed for a single pixel; for a neighbourhood of n pixels the following matrix form is obtained:

$$\begin{bmatrix} \frac{\partial I_{1_1}(x, y)}{\partial x} & \frac{\partial I_{1_1}(x, y)}{\partial y} \\ \vdots & \vdots \\ \frac{\partial I_{1_n}(x, y)}{\partial x} & \frac{\partial I_{1_n}(x, y)}{\partial y} \end{bmatrix} \begin{bmatrix} u \\ v \end{bmatrix} = \begin{bmatrix} I_{1_1}(x, y) - I_{2_1}(x, y) \\ \vdots \\ I_{1_n}(x, y) - I_{2_n}(x, y) \end{bmatrix}. \quad (3.16)$$

Equation (3.15) is similar to

$$A_1 x_1 = b_1. \quad (3.17)$$

The above system can be solved by using the pseudoinverse of A_1. First we need to multiply both sides of the equation by $(A_1)^T$ to obtain a quadratic matrix $A_1(A_1)^T$ on the left side of the equation:

$$A_1^T A_1 x_1 = A_1^T b_1. \quad (3.18)$$

Then multiply both sides by the inverse of $A_1(A_1)^T$ and the system is solved:

$$x_1 = (A_1^T A_1)^{-1} A_1^T b_1. \quad (3.19)$$

Therefore, we can express $(A_1^T A_1)$ as

$$\begin{bmatrix} \frac{\partial I_{1_1}}{\partial x} & \cdots & \frac{\partial I_{1_n}}{\partial x} \\ \frac{\partial I_{1_1}}{\partial y} & \cdots & \frac{\partial I_{1_n}}{\partial y} \end{bmatrix} \begin{bmatrix} \frac{\partial I_{1_1}}{\partial x} & \frac{\partial I_{1_1}}{\partial y} \\ \vdots & \vdots \\ \frac{\partial I_{1_n}}{\partial x} & \frac{\partial I_{1_n}}{\partial y} \end{bmatrix} = \begin{bmatrix} \sum_{i=1}^{n} \left(\frac{\partial I_{1_i}}{\partial x} \right)^2 & \sum_{i=1}^{n} \frac{\partial I_{1_i}}{\partial x} \frac{\partial I_{1_i}}{\partial y} \\ \sum_{i=1}^{n} \frac{\partial I_{1_i}}{\partial x} \frac{\partial I_{1_i}}{\partial y} & \sum_{i=1}^{n} \left(\frac{\partial I_{1_i}}{\partial y} \right)^2 \end{bmatrix}, \tag{3.20}$$

which has the same form as the first term of the right-hand side of equation (3.16). In addition, $(A_1)^T b_1$ yields

$$\begin{bmatrix} \frac{\partial I_{1_1}}{\partial x} & \cdots & \frac{\partial I_{1_n}}{\partial x} \\ \frac{\partial I_{1_1}}{\partial y} & \cdots & \frac{\partial I_{1_n}}{\partial y} \end{bmatrix} \begin{bmatrix} I_{1_1} - I_{2_1} \\ \vdots \\ I_{1_n} - I_{2_n} \end{bmatrix} = \begin{bmatrix} \sum_{i=1}^{n} \frac{\partial I_{1_i}}{\partial x} (I_{1_i} - I_{2_i}) \\ \sum_{i=1}^{n} \frac{\partial I_{1_i}}{\partial y} (I_{1_i} - I_{2_i}) \end{bmatrix}, \tag{3.21}$$

which corresponds to the second term of equation (3.18). The optical flow model is equivalent to the Lucas–Kanade algorithm. However, equation (3.18) has a double sum because the neighbourhood of the equation is $N \times M$ and the terms of equation (3.20) and equation (3.21) the neighbourhood are the same but use a one-dimensional path of length $n = N \times M$.

3.5 The Horn–Schunck algorithm

The Horn–Schunck method to estimate the optical flow is a global method that introduces a smoothness condition to solve the aperture problem. The method is defined as global because it determines the vector of movement of the optical flow in each pixel of the image, in contrast to local methods that only act on a few discrete coordinates chosen for their appearance (Gong and Bansmer 2015, González-Acuña *et al* 2019).

3.5.1 The smoothness principle

The smoothness in the optical flow refers to the transitions in areas of the images—a transition from a light to a dark area should occur smoothly, passing through the intermediate intensities. In the same way, the intensity change of a pixel in any coordinate to the corresponding one of the consecutive image must be smooth. The Horn–Schunck method is based on this requirement to calculate the derivatives of the intensities (in x and y space and in time) necessary to finally estimate the motion vectors of each pixel, which constitute the optical flow (Meinhardt-Llopis *et al* 2013).

3.5.2 The mathematical model

The Horn–Schunck algorithm assumes smoothness in the optical flow throughout the image. In this way it tries to minimize distortions in the flow and gives preference to the solutions of greater softness. This means that you are more likely to obtain sets of matching modulus and direction velocity vectors, which describe the motion of a large object, rather than having chaotic vectors indicating that each pixel is moving in a different direction.

The algorithm was developed to be applied to video. Video is a sequence of discrete images and each image is an array of discrete pixels. However, the foundation of the algorithm begins with a mathematical approach in continuous domains, with derivatives in space and time. After developing analytically in the continuous domain, in the last stage it is discretized to acquire its form for practical application.

The flow is formulated as global energy functional to minimize. It is a function of consecutive images of a video.

The Horn–Schunck method of estimating optical flow is global. The optical flow is formulated as a global functional, which is then aimed to be minimized. The functional is given by

$$\iint L dx dy, \tag{3.22}$$

where the Lagrangian is given by

$$L \equiv \left(\frac{\partial I_1(x, y)}{\partial x} u + \frac{\partial I_1(x, y)}{\partial y} v + \frac{\partial I(x, y)}{\partial t} \right)^2 + \phi^2 (||\nabla u||^2 + ||\nabla v||^2),$$

where

$$\frac{\partial I(x, y)}{\partial t} = I_2(x, y) - I_1(x, y) \tag{3.23}$$

and ϕ is a regularization constant. Larger values of ϕ lead to a smoother flow.

The functional of equation (3.23) is minimized by solving the Lagrangian equations in the original version, which are given by

$$\frac{\partial L}{\partial u} - \frac{\partial}{\partial x} \frac{\partial L}{\partial u_x} - \frac{\partial}{\partial y} \frac{\partial L}{\partial u_y} = 0 \tag{3.24}$$

and

$$\frac{\partial L}{\partial v} - \frac{\partial}{\partial x} \frac{\partial L}{\partial v_x} - \frac{\partial}{\partial y} \frac{\partial L}{\partial v_y} = 0. \tag{3.25}$$

Then we reduce the Lagrangian, taking account of the Laplace operator as

$$\Delta \equiv \frac{\partial^2}{\partial x^2} + \frac{\partial^2}{\partial y^2}. \tag{3.26}$$

However, we can numerically approximate $\Delta u(x, y)$ using finite differences, thus

$$\Delta u(x, y) = \bar{u}(x, y) - u(x, y), \tag{3.27}$$

where $\bar{u}(x, y)$ is a weighted average of u. Therefore, replacing equation (3.27) in equations (3.24) and (3.25) we have

$$0 = 2\frac{\partial I(x, y)}{\partial t}\frac{\partial I_1(x, y)}{\partial x} + 2\left[\frac{\partial I_1(x, y)}{\partial x}\right]^2 u + 2\frac{\partial I(x, y)}{\partial t}\frac{\partial I_1(x, y)}{\partial x}v + 2\phi^2[\bar{u}(x, y) - u(x, y)] \tag{3.28}$$

and

$$0 = 2\frac{\partial I(x, y)}{\partial t}\frac{\partial I_1(x, y)}{\partial x} + 2\left[\frac{\partial I_1(x, y)}{\partial x}\right]^2 u + 2\frac{\partial I(x, y)}{\partial t}\frac{\partial I_1(x, y)}{\partial x}v + 2\phi^2[\bar{u}(x, y) - u(x, y)]. \tag{3.29}$$

Solving this system for u and v and iterating until u and v converge at $k + 1$ iterations, we obtain that u_{k+1} and v_{k+1} are

$$u_{k+1} = \bar{u}(x, y) - \frac{\partial I_1(x, y)}{\partial x}\left[\frac{\frac{\partial I_1(x, y)}{\partial x}\bar{u}(x, y) + \frac{\partial I_1(x, y)}{\partial y}\bar{v}(x, y) + \frac{\partial I(x, y)}{\partial t}}{\phi^2 + \left(\frac{\partial I_1(x, y)}{\partial x}\right)^2 + \left(\frac{\partial I_1(x, y)}{\partial y}\right)^2}\right] \tag{3.30}$$

and in the vertical direction we have

$$v_{k+1} = \bar{v}(x, y) - \frac{\partial I_1(x, y)}{\partial y}\left[\frac{\frac{\partial I_1(x, y)}{\partial x}\bar{u}(x, y) + \frac{\partial I_1(x, y)}{\partial y}\bar{v}(x, y) + \frac{\partial I(x, y)}{\partial t}}{\phi^2 + \left(\frac{\partial I_1(x, y)}{\partial x}\right)^2 + \left(\frac{\partial I_1(x, y)}{\partial y}\right)^2}\right]. \tag{3.31}$$

The last expressions are the u_k and v_k displacements at iteration $k + 1$.

Among the advantages of the Horn–Schunck algorithm the high density of flow vectors stands out—flow information that is lost in homogeneous areas is padded from the edges with flow information. Among the disadvantages is that it is more sensitive to noise than local methods such as the Lucas–Kanade algorithm. The `Python` code of the Horn–Schunck algorithm is left as an exercise for the reader.

3.6 End notes

In this chapter we have seen the two main algorithms of optical flow: the Lucas–Kanade algorithm and the Horn–Schunck algorithm. The Lucas–Kanade algorithm works by minimizing the error via the least-squares criterion. In the Lucas–Kanade algorithm a system of equations is solved locally. The Horn–Schunck algorithm works by reducing a global functional. The computations of the Horn–Schunck algorithm are global rather than local.

References

Baker S and Matthews I 2004 Lucas–Kanade 20 years on: a unifying framework *Int. J. Comput. Vis.* **56** 221–55

Baker S, Gross R, Ishikawa T and Matthews I 2003 Lucas–Kanade 20 years on: a unifying framework: Part 2 *Int. J. Comput. Vis.* **56** 221–55

Bauer N, Pathirana P and Hodgson P 2006 Robust optical flow with combined Lucas–Kanade/Horn–Schunck and automatic neighborhood selection *2006 Intr. Conf. on Information and Automation* (Piscataway, NJ: IEEE) pp 378–83

Beauchemin S S and Barron J L 1995 The computation of optical flow *ACM Comput. Surv.* **27** 433–66

Bouguet J-Y *et al* 2001 Pyramidal implementation of the affine Lucas–Kanade feature tracker description of the algorithm *Intel Corp.* **5** 4

Bruhn A, Weickert J and Schnörr C 2005 Lucas/Kanade meets Horn/Schunck: combining local and global optic flow methods *Int. J. Comput. Vis.* **61** 211–31

Fleet D and Weiss Y 2006 Optical flow estimation *Handbook of Mathematical Models in Computer Vision* (Berlin: Springer) pp 237–57

Gong X and Bansmer S 2015 Horn–Schunck optical flow applied to deformation measurement of a birdlike airfoil *Chin. J. Aeronaut.* **28** 1305–15

González-Acuña R G, Dávila A and Gutiérrez-Vega J C 2019 Optical flow of non-integer order in particle image velocimetry techniques *Signal Process.* **155** 317–22

Grant I 1997 Particle image velocimetry: a review *Proc. Inst. Mech. Eng.* C **211** 55–76

Meinhardt-Llopis E, Pérez J S and Kondermann D 2013 Horn–Schunck optical flow with a multi-scale strategy *Image Process. Online* **3** 151–72

Raffel M, Willert C E, Scarano F, Kähler C J, Wereley S T and Kompenhans J 2018 *Particle Image Velocimetry: A Practical Guide* (Berlin: Springer)

Rav-Acha A and Peleg S 2006 Lucas–Kanade without iterative warping *2006 Intr. Conf. on Image Processing* (Piscataway, NJ: IEEE) pp 1097–100

Santiago J G, Wereley S T, Meinhart C D, Beebe D J and Adrian R J 1998 A particle image velocimetry system for microfluidics *Exp. Fluids* **25** 316–9

Schalkoff R J 1989 *Digital Image Processing and Computer Vision* vol 286 (New York: Wiley)

Schreiber D 2008 Generalizing the Lucas–Kanade algorithm for histogram-based tracking *Pattern Recognit. Lett.* **29** 852–61

Tong W 2013 Formulation of Lucas–Kanade digital image correlation algorithms for non-contact deformation measurements: a review *Strain* **49** 313–34

Westerweel J 1997 Fundamentals of digital particle image velocimetry *Meas. Sci. Technol.* **8** 1379

Willert C E and Gharib M 1991 Digital particle image velocimetry *Exp. Fluids* **10** 181–93

Yegavian R, Leclaire B, Champagnat F, Illoul C and Losfeld G 2016 Lucas–Kanade fluid trajectories for time-resolved PIV *Meas. Sci. Technol.* **27** 084004

Rafael G González-Acuña, Héctor A Chaparro-Romo and Israel Melendez-Montoya

Chapter 4

Object detection algorithms

In this chapter we study the concept of object detectors. We start from the most straightforward template matching and move on to trained descriptors.

4.1 Object detection

In many computer vision applications the goal is to find a particular object in an image. To do this a series of steps are implemented through an object detection algorithm (Chityala and Pudipeddi 2020, Forsyth and Ponce 2002, Raschka 2015, Schalkoff 1989).

The problem lies in the generality of the problem, for example, the object can be anything and the image can have an arbitrary number of objects. There are several ways to solve the problem. For particular objects in specific images, some algorithms may perform better than others.

In this chapter we will first study different ways to solve the problem of detecting objects in images. We start with template matching, which is a technique in digital image processing for finding small parts of an image which match a template image.

Template matching compares a template against overlapped image regions. The comparison can be achieved using several methods. In this chapter we are going to study the correlation method, which is one of the most popular. The other methods, such as the least-squares difference, normalized least-squares difference or normalized correlation, are very similar.

4.1.1 Statistical interpretation of correlation

Rather than estimating the matching between images analytically, the method of choice is to find the best match locally between the subimage I_{s1} and the subimage I_{s2} in a statistical way. In the introduction of this chapter we mentioned that it is complex to achieve template matching. Thus the analytical estimation is avoided and is replaced by a statistical computation. Estimating the template matching can be achieved through the use of the discrete cross-correlation function.

doi:10.1088/978-0-7503-3707-6ch4

To understand the statistical background of the cross-correlation function,

$$R(x, y) = \sum_{i=-k}^{k} \sum_{i=-l}^{l} I_{s1}(i, j) I_{s2}(i + x, j + y). \tag{4.1}$$

$R(x, y)$ is the measurement of correlation of the two samples at the current pixel of index (x, y). The size of the neighbourhood is determined by k and l. The variables I_{s1} and I_{s2} are the subimages; one of them is the template. I_{s2} is more extensive than I_{s1}essentially the template I_{s1} is linearly shifted around the sample I_{s2} without extending over the edges of I_{s2}.

This expression is called the discrete cross-correlation. The cross-correlation function measures statistically the degree of matching between the two samples for a given shift. The highest value in the correlation plane can be used as a direct estimate of the template matching (see figure 4.1).

The cross-correlation will give a peak in the correlation plane when the particle in subimage I_{s1} finds itself in subimage I_{s2}. This happens because the cross-correlation is just a sum of the products of the intensities between the two subimages. Determining the coordinates of the peak yields the template matching. The correlation also has a Fourier interpretation.

4.1.2 Fourier interpretation of correlation

The Fourier transform has many useful properties, such as linearity, time-shifting, scaling factor, etc. However, there is a crucial property of the Fourier transform for our purposes, the correlation property. To explain it we first need to recall what the Fourier transform is.

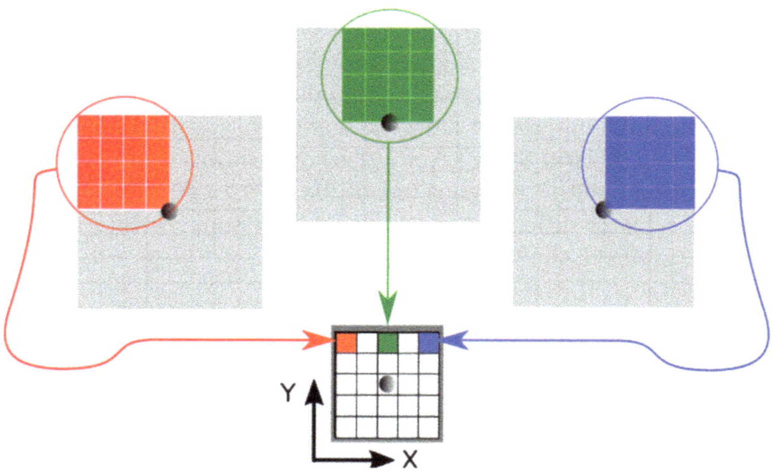

Figure 4.1. Example of the formation of the correlation plane by direct cross-correlation. Here a 4 × 4 px sample is correlated with a larger 8 × 8 px sample to produce a 5 × 5 px correlation plane.

The following equation defines the Fourier transform:

$$F(\omega) = \int_{-\infty}^{+\infty} f(x)e^{-2\pi i\omega x}dx, \tag{4.2}$$

where $f(x)$ is the function to be transformed, $F(\omega)$ is the transformation of the function $f(x)$, the exponential is the kernel of the Fourier transform, i is the imaginary number, x is the spatial variable and ω is the variable of frequency.

Thus the Fourier transform of a function $f(x)$ is just the decomposition of the signal in its frequency components. The Fourier transform is just the Fourier complex series applied to a function that has an infinity period.

There is an anti-transformation operation, which has the following form:

$$f(x) = \int_{-\infty}^{+\infty} F(\omega)e^{2\pi i\omega x}d\omega. \tag{4.3}$$

On the other side, the convolution in one dimension is defined as

$$f(x) * g(x) = \int_{-\infty}^{+\infty} f(x - x')g(x')dx'. \tag{4.4}$$

The correlation is almost the same as the convolution, but it has a difference, a minus sign. This difference creates a different concept. The correlation is represented in equation (4.1):

$$f(x) \circ g(x) = \int_{-\infty}^{+\infty} f(x + x')g(x')dx'. \tag{4.5}$$

Now when we apply the Fourier transform to the correlation, we will prove that the product of the Fourier transform gives the transform results of both functions in two dimensions.

First we define the two functions to have the following forms in terms of their Fourier transform, respectively:

$$f(x, y) = \mathscr{F}^{-1}[F(\omega_x, \omega_y)] = \int_{-\infty}^{+\infty}\int_{-\infty}^{+\infty} F(\omega_x, \omega_y)e^{2\pi i(x\omega_x + y\omega_y)}d\omega_x d\omega_y \tag{4.6}$$

and

$$g(x, y) = \mathscr{F}^{-1}[G(\omega_x, \omega_y)] = \int_{-\infty}^{+\infty}\int_{-\infty}^{+\infty} G(\omega_x, \omega_y)e^{2\pi i(x\omega_x + y\omega_y)}d\omega_x d\omega_y. \tag{4.7}$$

On the other hand, the correlation for these functions of two dimensions has the following form:

$$f(x, y) \circ g(x, y) = \int_{-\infty}^{+\infty}\int_{-\infty}^{+\infty} f(x + x', y + y')g(x', y')dx'dy'. \tag{4.8}$$

Replacing equation (4.6) in equation (4.8) we obtain

$$f(x, y)\circ g(x, y) = \int_{-\infty}^{+\infty}\int_{-\infty}^{+\infty} g(x', y') \times \int_{-\infty}^{+\infty}\int_{-\infty}^{+\infty} F(\omega_x, \omega_y)e^{2\pi i[\omega_x(x+x')+\omega_y(y+y')]} \times dw_x dw_y dx' dy'. \tag{4.9}$$

Reordering equation (4.9) we have

$$f(x, y)\circ g(x, y) = \int_{-\infty}^{+\infty}\int_{-\infty}^{+\infty} F(\omega_x, \omega_y)dw_x dw_y \times \int_{-\infty}^{+\infty}\int_{-\infty}^{+\infty} g(x', y')e^{2\pi i[\omega_x(x+x')+\omega_y(y+y')]} \times dx' dy'. \tag{4.10}$$

In equation (4.10) the kernel has a positive sign; the integration inside the brackets will give the conjugate of the Fourier transform of $g(x', y')$:

$$f(x, y)\circ g(x, y) = \int_{-\infty}^{+\infty}\int_{-\infty}^{+\infty} F(\omega_x, \omega_y)G^*(\omega_x, \omega_y)e^{2\pi i[w_x(x)+w_y(y)]}dw_x dw_y. \tag{4.11}$$

Equation (4.11) can be represented in the following form:

$$f(x, y)\circ g(x, y) = \mathscr{F}^{-1}[F(\omega_x, \omega_y)G^*(\omega_x, \omega_y)]. \tag{4.12}$$

Therefore, the transform of correlation of the images $g(x, y)$ and $f(x, y)$ is

$$\mathscr{F}[f(x, y)\circ g(x, y)] = F(\omega_x, \omega_y)G^*(\omega_x, \omega_y). \tag{4.13}$$

The transform of the correlation of the functions $g(x, y)$ and $f(x, y)$ is just the multiplication of the Fourier transform of the function $f(x, y)$ with the conjugate of the Fourier transform of the function $g(x, y)$.

This demonstration is for continuous data, but if integrals are by replaced by summations it can be applied to discrete data as well. Then this equation can be used for our purposes of template matching.

In `OpenCV` template matching is already implemented. In the algorithm of listing 4.1 in `Python` we use the function `cv2.matchTemplate` to compute template matching between two images (see figure 4.2 for the output).

Note the documentation of `cv2.matchTemplate`. Other methods of correlation and matching are available, such as least-squares difference, normalized least-squares difference and normalized correlation, which are very similar.

However, the truth is that template matching is not a robust algorithm and is prone to false positives. Thus all of this discussion has served as an introduction to the algorithms that we will study in the following sections.

4.2 Sliding windows and image pyramids

Before we continue with the goal of developing powerful algorithms of template matching, we first need to focus on two concepts: sliding windows and image pyramids. A sliding window is a rectangular region that crosses over the image (Gonzalez *et al* 2004, Szeliski 2010).

```python
import cv2 as ocv  # load opencv

# load the image and the template
image =ocv.imread("image1.jpg" )
temp = ocv.imread("image.png" )
(H,W) = temp.shape[:2]

# find the template in the source image
result = ocv.matchTemplate(image,temp,ocv.TM_CCOEFF)
(minVal,maxVal,minLoc,(x,y))=ocv.minMaxLoc(result)

# draw the bounding box on the source image
ocv.rectangle(image,(x,y),(x+W,y+H),(0,0,255),2)

# show results
ocv.imshow("Image␣with␣the␣match␣",image)
ocv.imshow("Template␣of␣the␣matching",temp)
ocv.waitKey(0)
```

Listing 4.1 The matching template.

Figure 4.2. Output for listing 4.1: the matching template. Credit: Highsmith, Carol M., 1946 - United States Library of Congress's Prints and Photographs Division under the digital ID highsm.09920; and Jerry Stratton / http://hoboes.com/Mimsy/CC BY-SA 3.0.

The rectangular region has a fixed width and height. The code in listing 4.2 is a function in `Python` for a sliding window (see figure 4.3 for the ouput). Image pyramids are multi-scale representations of images (see figure 4.4).

This means that the original image is represented with different sizes. Thus to obtain the image pyramids of an image we need to use the function `imutils.resize` on the image several times (Kapur 2017, Lutz 2001, Oliphant 2007, Umesh 2012).

```
#from pyimagesearch.object_detection.helpers import pyramid
import argparse
import cv2
import imutils #load imutils

def pyramid(image, scale=1.5, MinImageSize=(30, 30)):
        # yield image
        yield image

        # while is true get pyramids
        while True:

                w = int(image.shape[1] / scale)
                image = imutils.resize(image, width=w)

                if image.shape[0] < MinImageSize[1] or
                   image.shape[1] < MinImageSize[0]:
                        break

                yield image

# construct the argument parser and parse the arguments
ap = argparse.ArgumentParser()
ap.add_argument("-i", "--image", required=True, help="path␣
   to␣the␣input␣image")
ap.add_argument("-s", "--scale", type=float, default=1.5,
   help="scale␣factor␣size")
args = vars(ap.parse_args())

image = cv2.imread(args["image"])
# loop over the layers of the image pyramid and display
   them
for(i, layer) in enumerate(pyramid(image, scale=args["scale
   "])):
        cv2.imshow("Layer␣{}".format(i + 1),layer)
        cv2.waitKey(0)
```

Listing 4.2 The sliding pyramid operation.

4.3 The histogram of oriented gradients descriptor

The histogram of oriented gradients (HOG) is a feature descriptor used in artificial vision. The goal of the HOG is object detection. The HOG counts the appearances of a directed gradient in localized pieces of an image. In this section we show how to compute the HOG over an image step by step (Granlund and Knutsson 2013, Klette 2014).

Since HOG works with gradients, it is necessary to recall the gradients of images from chapter 2. Recall that the derivative of image I in x is given by

$$G_x = I^*\begin{bmatrix} -1 & 0 & +1 \\ -2 & 0 & +2 \\ -1 & 0 & +1 \end{bmatrix}. \tag{2.6}$$

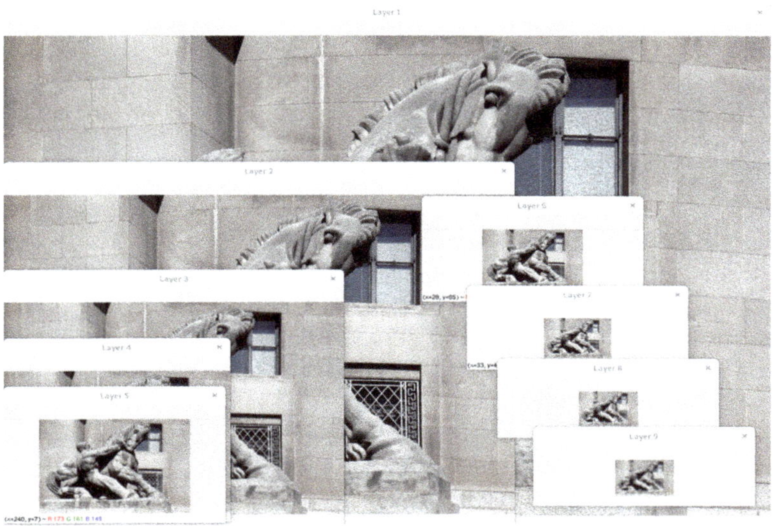

Figure 4.3. Output for listing 4.2: sliding pyramid operation. Credit: Highsmith, Carol M., 1946 - United States Library of Congress's Prints and Photographs Division under the digital ID highsm.09920.

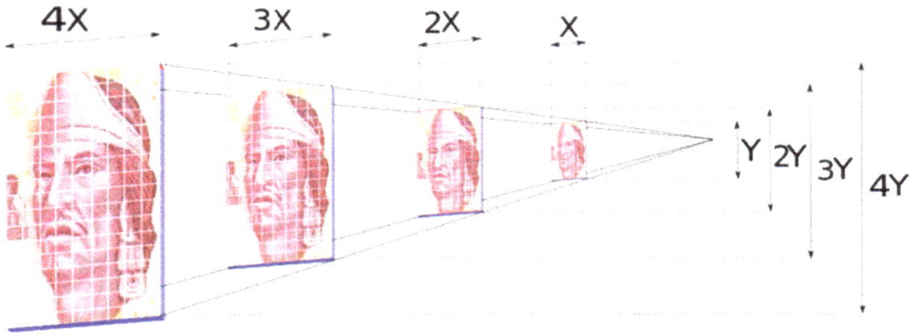

Figure 4.4. Pyramid building of an image scaled.

The derivative of image I in y is given by

$$G_y = I * \begin{bmatrix} -1 & -2 & -1 \\ 0 & 0 & 0 \\ +1 & +2 & +1 \end{bmatrix}. \tag{2.7}$$

We recall the gradient from

$$G = \sqrt{G_x + G_y} \tag{2.8}$$

and its direction from

$$\Theta = \arctan\left(\frac{G_x}{G_y}\right). \tag{2.9}$$

Thus the first step is to compute the gradient of the image. The second step is to compute the direction of the gradient of the image.

Then the image is divided into several subimages. All the subimages must have the same size. Let us say a subimage has $w \times h$ pixels, where w is the width in pixels and h is the height in pixels.

For each subimage of I we need to compute the histogram. This histogram takes a bin, which is selected based on the direction. The vote (the value that goes into the bin) is selected based on the magnitude.

Since we compute the direction with θ, the bin has its possible values. Commonly, the bin is taken from 0 to 180, in ranges of 20. Thus in this way the possible values of the bin are [0, 20, 40, 60, 80, 100, 120, 140, 160]. If the angle is greater than 160 degrees, it is between 160 and 180, and we know the angle wraps around making 0 and 180 equivalent.

The contributions of all the pixels in the subimage of size $n \times m$ are added up to create the nine bin histogram. Then we simply normalize the 9×1 histogram of each cell/subimage. Alternatively, we can normalize with blocks. A block is also a subimage but with a different size to the size implemented to compute the oriented histogram.

Thus if we let the block have the same size as the cell/subimage, the normalization is divided into the values of the bin i,

$$v_i \rightarrow \frac{v_i}{\sqrt{v_0^2 + v_{20}^2 + v_{40}^2 + v_{60}^2 + v_{80}^2 + v_{100}^2 + v_{120}^2 + v_{160}^2}}, \tag{4.14}$$

where v_i is the value of the bin i, $i \in \{0, 20, 40, 60, 80, 100, 120, 140, 160\}$. Finally, all the normalized values of the histograms of all the subimages are placed in a single vector.

OpenCV has the HOG algorithm with the function called cv2.HOGDescriptor. The code in listing 4.3 shows its functionality (see figure 4.5 for the output).

```
from skimage import exposure
from skimage import feature
import cv2

image = cv2.imread("image.jpg")
logo = cv2.resize(image, (640, 480))
(H, hogImage) = feature.hog(logo, orientations=9,
   pixels_per_cell=(8, 8),
         cells_per_block=(2, 2), transform_sqrt=True,
            block_norm="L1",
         visualize=True)
hogImage = exposure.rescale_intensity(hogImage, out_range
   =(0, 255))
hogImage = hogImage.astype("uint8")

cv2.imshow("HOG Image", hogImage)
cv2.waitKey(0)
```

Listing 4.3. The HOG descriptor.

Original Image　　　　　Second Original Image

Figure 4.5. Output for listing 4.3: the HOG descriptor. Credit: Highsmith, Carol M., 1946 - United States Library of Congress's Prints and Photographs Division under the digital ID highsm.09920.

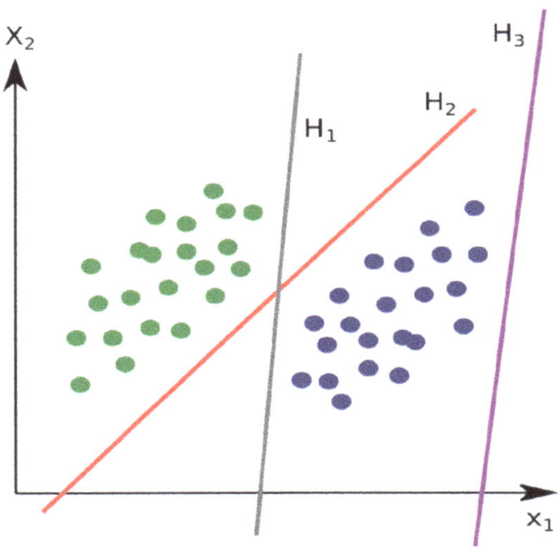

Figure 4.6. Schematic of the support vector machine.

4.4 Support vector machine

Support vector machines or SVMs are a set of supervised learning algorithms developed by Vladimir Vapnik. These methods are correctly related to classification and regression problems. Given a set of training examples (samples), we can label the classes and train an SVM to build a model that predicts the class of a new sample. Intuitively, an SVM is a model that represents the sample points in space, separating the classes into two spaces as wide as possible by means of a separation hyperplane defined as the vector between the two points of the two classes closest to them, which is called a support vector (see figure 4.6). When the new samples are put into correspondence with the said model, depending on the spaces to which they belong, they can be classified into one or the other class.

More formally, an SVM constructs a hyperplane or set of hyperplanes in a very high (or even infinite) dimensional space that can be used in classification or regression problems. A good separation between the classes will allow a correct classification.

4.4.1 The concepts behind the SVM

Given a set of points, a subset of a more extensive set, in which each of them belongs to one of two possible categories, an algorithm based on an SVM builds a model capable of predicting whether a new point, whose category we do not know, belongs to one category or the other. As in most supervised classification methods, the input data, the points, are viewed as a p-dimensional vector, an ordered list of p numbers.

The SVM looks for a hyperplane that optimally separates the points of one class from those of another, which eventually could have been previously projected into a space of higher dimensionality. In this concept of *optimal separation* is where the fundamental characteristic of SVM resides: this type of algorithm looks for the hyperplane that has the maximum distance with the points that are closest to itself. This is also why SVMs are sometimes referred to as top margin classifiers. In this way the vector points that are labelled with one category will be on one side of the hyperplane and the cases that are in the other category will be on the other side.

Figure 4.6 shows an idealized example for two dimensions, where the representation of the data to be classified is performed in the x–y plane. The SVM algorithm tries to find a one-dimensional hyperplane that joins the predictor variables and constitutes the limit that defines whether an input element belongs to one category or the other.

There is an infinite number of possible hyperplanes that perform the classification, but which one is the best and how do we define it? The best solution is the one that allows a maximum margin between the elements of the two categories. In chapter 8 we are going to see examples of SVM in `Python`.

4.5 End notes

In this chapter we introduce the problem of template matching. We introduce helpful concepts such as sliding windows and image pyramids. Finally, we study the HOG descriptor in detail. In the next chapter we focus on other important descriptors.

References

Chityala R and Pudipeddi S 2020 *Image Processing and Acquisition Using* `Python` (Boca Raton, FL: CRC Press)

Forsyth D A and Ponce J 2002 *Computer Vision: A Modern Approach* (Englewood Cliffs, NJ: Prentice Hall)

Gonzalez R C, Woods R E and Eddins S L 2004 *Digital Image Processing Using MATLAB* (New Delhi: Pearson Education India)

Granlund G H and Knutsson H 2013 *Signal Processing for Computer Vision* (Berlin: Springer)

Kapur S 2017 *Computer Vision with* `Python` *3* (Birmingham: Packt)

Klette R 2014 *Concise Computer Vision* (Berlin: Springer)

Lutz M 2001 *Programming* `Python` (Sebastopol, CA: O'Reilly)

Oliphant T E 2007 Python for scientific computing *Comput. Sci. Eng.* **9** 10–20
Raschka S 2015 *Python Machine Learning* (Birmingham: Packt)
Schalkoff R J 1989 *Digital Image Processing and Computer Vision* vol 286 (New York: Wiley)
Szeliski R 2010 *Computer Vision: Algorithms and Applications* (Berlin: Springer)
Umesh P 2012 Image processing in `Python` *CSI Communications* **23** 2

IOP Publishing

Optics and Artificial Vision

Rafael G González-Acuña, Héctor A Chaparro-Romo and Israel Melendez-Montoya

Chapter 5

Image descriptors

In this chapter we study the main feature descriptor algorithms applied in artificial vision. We begin with the basic statistical information of the image and move on to the keypoint detectors, local invariant descriptors and binary descriptors.

5.1 Introduction to image descriptors

In the previous chapter we studied an object detector algorithm, which is different to image descriptors. A feature detector is an algorithm which takes an image and outputs locations (i.e. pixel coordinates) of significant areas in your image. An example of this is a corner detector, which outputs the locations of corners in your image but does not tell you any other information about the features detected. When it comes to feature detectors, the *location* might also include a number describing the size or scale of the feature (Gonzalez *et al* 2004, Granlund and Knutsson 2013, Klette 2014, Schalkoff 1989).

A feature descriptor is an algorithm which takes an image and outputs feature descriptors/feature vectors. Feature descriptors encode interesting information into a series of numbers and act as a numerical *fingerprint* that can be used to differentiate one feature from another. Ideally, this information would be invariant under image transformation, so we can find the feature again even if the image is transformed in some way. An example would be SIFT, which encodes information about the local neighbourhood image gradients in the numbers of the feature vector. The numerical *fingerprint* of the image is located in feature vectors. Thus feature vectors are simply the unit of memory to hold the numerical description of the image (Chityala and Pudipeddi 2020, Forsyth and Ponce 2002).

In this chapter we are going to study the main feature descriptor algorithms applied in artificial vision. We are going to start with the most basic descriptors and finish with some of the more robust descriptor algorithms.

doi:10.1088/978-0-7503-3707-6ch5

5.2 Basic statistics

The basic image properties of an image are the basic statistical properties: the mean value and the standard deviation (Hu 1962). To obtain the mean value and the standard deviation in `OpenCV` we use the command `cv2.meanStdDev`. The code in listing 5.1 computes the mean value and the standard deviation of a set of images and locates the results in the feature vector (see figure 5.1 for the output).

The command `Path[Path.rfind("/" ) + 1:]` takes the name of the file. For example, if the `Path` is "`Desktop/Documents/Images/Dog.png`", then `filename` will be "`Dog.png`" (Lutz 2001, Umesh 2012, Van Rossum and Drake 1995).

5.3 Hu moments

The following descriptors that we are going to study are the Hu moments. Hu moments are applied to describe binary images. Recall that binary images are the images with pixel values of 0 or 255, black or white. Hu moments, or rather Hu moment invariants, are a set of seven numbers calculated using moments that are invariant under image transformations.

Moments are statistical expectations of a random variable on a probability distribution. The idea behind Hu moments is that we can see binary images as probability distributions. Then we compute the moments and from them we can obtain invariants. Invariants are quantities that do not change under image transformations, such as translations, scales, rotations and reflections. Invariants are what we want—a description of the image that does not change. A classifying invariant can be used to fully categorize images. Thus let us start our study of Hu moments with the regular moment of a shape in a binary image, which is given by

$$M_{i,j} = \sum_{x} \sum_{y} x^i y^j I(x, y), \tag{5.1}$$

```python
import cv2 as ocv
import imutils

image = ocv.imread("image.jpg")
image = ocv.cvtColor(image, ocv.COLOR_BGR2GRAY)

(means, stds) = ocv.meanStdDev(image)
print("Means: ",means)
print("Standard Desviation:",stds)
ocv.waitKey(0)
```

Listing 5.1 Mean and standard deviation operations.

```
$ python listing_5.1.py
('Means: ', array([[154.85885251]]))
('Standard Desviation:', array([[58.56402817]]))
```

Figure 5.1. Output of listing 5.1: mean and standard deviation operation.

where $\sum_x \sum_y$ is a sum over the whole image and $I(x, y)$ is the intensity value of the pixel at (x, y). From equation (5.1) we can obtain the centre of the shape of the binary image. In the x direction we have

$$\bar{x} = \frac{M_{10}}{M_{00}} \tag{5.2}$$

and in the y direction we have

$$\bar{y} = \frac{M_{01}}{M_{00}}. \tag{5.3}$$

The coordinates $(\bar{x}, \bar{y})$ are called the centroid coordinates because they indicate where the centre of the object of the binary image is. From the centroid we can compute the relative moments of the image, which are given by

$$\mu_{pq} = \sum_x \sum_y (x - \bar{x})^i (y - \bar{y})^j I(x, y). \tag{5.4}$$

As mentioned above, Hu moments are moments construed by relative moments such that they are invariant under translations, scales, rotations and reflections.

The first Hu moment is

$$M_1 = \mu_{20} + \mu_{02}. \tag{5.5}$$

The second Hu moment has terms of μ_{20}, μ_{02} and μ_{11}, it is given by

$$M_2 = (\mu_{20} - \mu_{02})^2 + 4\mu_{11}^2. \tag{5.6}$$

The third Hu moment is denoted by

$$M_3 = (\mu_{30} - 3\mu_{12})^2 + (3\mu_{21} - \mu_{30})^2. \tag{5.7}$$

The fourth Hu moment also depends on μ_{30}, μ_{03}, μ_{21} and μ_{12},

$$M_4 = (\mu_{30} + \mu_{12})^2 + (\mu_{21} + \mu_{30})^2. \tag{5.8}$$

The fifth Hu moment is given by the following expression:

$$\begin{aligned} M_5 = &(\mu_{30} - 3\mu_{12})(\mu_{30} + \mu_{12})[(\mu_{30} + \mu_{12})^2 - 3(\mu_{21} + \mu_{03})^2] \\ &+ (3\mu_{21} - \mu_{03})(\mu_{21} + \mu_{03})[3(\mu_{30} + \mu_{12})^3 - (\mu_{21} + \mu_{03})^2]. \end{aligned} \tag{5.9}$$

The sixth term is

$$M_6 = (\mu_{20} - \mu_{02})[(\mu_{30} + \mu_{12})^2 - (\mu_{21} + \mu_{03})^2] + 4\mu_{11}(\mu_{30} + 3\mu_{12})(\mu_{21} + \mu_{03}). \tag{5.10}$$

Finally, the last term of the Hu moments, the seventh moment, is

$$\begin{aligned} M_7 = &(3\mu_{21} - \mu_{03})(\mu_{30} + \mu_{12})[(\mu_{30} + \mu_{12})^2 - 3(\mu_{21} + \mu_{03})^2] \\ &- (\mu_{30} - 3\mu_{12})(\mu_{21} + \mu_{03})[3(\mu_{30} + \mu_{12})^3 - (\mu_{21} + \mu_{03})^2]. \end{aligned} \tag{5.11}$$

```
import cv2 as ocv
import imutils

image = ocv.imread("original.jpg")
image = ocv.cvtColor(image, ocv.COLOR_BGR2GRAY)

# compute Hu Moments and add them in a feature vector
moments = ocv.HuMoments(ocv.moments(image)).flatten()
print("Hu moments: {}".format(moments))
ocv.waitKey(0)
```

Listing 5.2 Hu moments operation.

```
$ python listing_5.2.py
Hu moments: [ 1.21196105e-03  1.40884559e-07  4.89960090e-13  7.90095796e-13
  1.06037118e-26  5.95777693e-18 -4.91472159e-25]
```

Figure 5.2. Output of listing 5.2: Hu moments operation.

In the code in listing 5.2 the Hu moments are computed by the command `cv2.HuMoments` (see figure 5.2 for the output).

5.4 Zernike moments

The next moments we want to study are the Zernike moments, which, like Hu moments, also describe the shape of an object in an image. Zernike moments characterize the shape of an object in a binary image. This object can be a contour, a boundary or a solid object.

Zernike moments are based on the theory of orthogonal functions. Two functions f and g in a certain space are said to be orthogonal if their dot product $\langle f, g\rangle$ is null.

Whether two particular functions are orthogonal depends on how their dot product has been defined. A very common definition of a dot product between functions is

$$\langle f, g\rangle = \int_a^b f^*(x)g(x)w(x)dx, \tag{5.12}$$

with appropriate limits of integration and where * denotes a complex conjugate and $w(x)$ is a weight function. In many applications we take $w(x) = 1$. Examples of orthogonal functions are the sine and cosine function with $w(x) = 1$.

The value of orthogonal functions is that there is no repetition of data connecting moments, making them more robust and discriminative than Hu moments, which are based on simple mathematical roots.

Zernike moments were first introduced in the research paper 'Image analysis via general theory of moments'. In appendix A of that paper the Zernike moments are expressed in terms of the usual moments and the following expressions are the Zernike moments (Teague 1980).

The second-order Zernike moments are

$$S_1 = \frac{3[2(\mu_{20} + \mu_{02}) - 1]}{\pi} \tag{5.13}$$

and

$$S_2 = \frac{9[(\mu_{20} + \mu_{02})^2 - 4(\mu_{11})^2]}{\pi^2}. \tag{5.14}$$

The third-order Zernike moments are given by

$$S_3 = \frac{16[(\mu_{30} - 3\mu_{21})^2 + (\mu_{30} - 3\mu_{12})^2]}{\pi^2}, \tag{5.15}$$

the fourth Zernike moment by

$$S_4 = \frac{144[(\mu_{03} + \mu_{21})^2 + (\mu_{30} + \mu_{12})^2]}{\pi^2}, \tag{5.16}$$

the fifth Zernike moment by

$$\begin{aligned} S_5 = \frac{13\,824}{\pi^4}\{&(\mu_{03} - 3\mu_{21})(\mu_{03} + \mu_{21})[(\mu_{03} + \mu_{21})^2 - 3(\mu_{30} + \mu_{12})^2] \\ &- (\mu_{30} - 3\mu_{12})(\mu_{30} + \mu_{12})[(\mu_{30} + \mu_{12})^2 - 3(\mu_{30} + \mu_{12})^2]\} \end{aligned} \tag{5.17}$$

and the sixth Zernike moment is given by

$$\begin{aligned} S_6 = \frac{864}{\pi^3}\{&(\mu_{02} - \mu_{20})[(\mu_{03} + \mu_{21})^2 - (\mu_{30} + \mu_{12})^2] + 4\mu_{11}(\mu_{03} + \mu_{21}) \\ &(\mu_{30} + \mu_{12})\}. \end{aligned} \tag{5.18}$$

The fourth-order Zernike moments are given by

$$S_7 = \frac{25[(\mu_{40} - 6\mu_{22} + \mu_{04})^2 + 16(\mu_{31} - \mu_{13})^2]}{\pi}, \tag{5.19}$$

$$S_8 = \frac{25\{4(\mu_{04} - \mu_{40}) + 3(\mu_{20} - \mu_{02})^2 + 4[4(\mu_{31} + \mu_{13}) - 3\mu_{11}]\}^2}{\pi}, \tag{5.20}$$

$$S_9 = \frac{5[6(\mu_{40} + 2\mu_{22} + \mu_{04}) - 6(\mu_{20} + \mu_{02}) + 1]}{\pi}, \tag{5.21}$$

$$\begin{aligned} S_{10} = \frac{250}{\pi^3}(&(\mu_{40} - 6\mu_{22} + \mu_{04})\{4(\mu_{04} - \mu_{40}) + 3(\mu_{20} - \mu_{02})^2 - 4[4(\mu_{31} + \mu_{13}) - 3\mu_{11}]^2\} \\ &- 16[4(\mu_{04} - \mu_{40}) + 3(\mu_{20} - \mu_{02})][4(\mu_{31} + \mu_{13}) - 3\mu_{11}](\mu_{31} - \mu_{13})) \end{aligned} \tag{5.22}$$

```
# load libraries
import cv2 as ocv
import imutils
import mahotas
import mahotas . demos
from pylab import gray , imshow , show
import numpy as np
import matplotlib . pyplot as plot
# load iamge
image = ocv.imread("image.jpg")
ocv.imshow("Image", image)
# filtering image
img = image.max(2)
# showing image
ocv.imshow("Image␣process", img)
ocv.waitKey()
# radius
radius = 10
# compute zernike moments
value = mahotas.features.zernike_moments(img,radius)
# print zernike moments
print(value)
```

Listing 5.3 Zernike moments operation.

```
$ python listing_5.3.py
[0.31830989 0.01034137 0.00586077 0.00264844 0.01443975 0.00841975
 0.00249042 0.00752401 0.0088511  0.01907787 0.01358087 0.00227515
 0.01108504 0.00809162 0.00766836 0.00616902 0.02516596 0.02337153
 0.00410642 0.03376145 0.02180296 0.01300231 0.00868495 0.01107335
 0.00222286]
```

Figure 5.3. Output of listing 5.3: Zernike moments operation.

and

$$S_{11} = \frac{30}{\pi^2}\{[4(\mu_{04} - \mu_{40}) + 3(\mu_{20} - \mu_{02})](\mu_{02} - \mu_{20}) + 4\mu_{11}[4(\mu_{31} + \mu_{13}) - 3\mu_{11}]\}. \quad (5.23)$$

Zernike moments are computed in `Python` with a command from the `mahotas` library, and the command is `mahotas.features.zernike_moments` (see listing 5.3 for the code and figure 5.3 for the output).

5.5 Haralick features

Haralick features are descriptors used to characterize the texture of an image. The application of Haralick features distinguishes between rough and smooth surfaces.

Robert M Haralick first proposed these features in 1973, in his paper 'Textural features for image classification' (Haralick *et al* 1973).

In order to analyse the textures of an image, Haralick features are based on a co-occurrence matrix. A co-occurrence matrix is a matrix that is defined over an image to measure co-occurring pixel values. This matrix can work with pixels of greyscale values, or colours. The co-occurring matrix uses a given offset to measure the co-occurring pixel values.

Now let us compute the co-occurrence matrix for a grey level image I. This means that we compute pairs of pixels with a specific value and offset occurs in the image. The offset $(\delta x, \delta y)$ is a position operator that can be applied to any pixel. δx can be seen as the horizontal displacement and δy can be seen as the vertical displacement of pixel $I(x, y)$. Thus the displacement leads us to $I(x + \delta x, y + \delta y)$.

Therefore an image with p different pixel values will produce a co-occurrence matrix with $p \times p$ different values, for $(\delta x, \delta y)$. The value at (i, j) in the co-occurrence matrix is the number of times when i and j pixel values repeat with respect to the defined offset used.

Thus let C be the co-occurrence matrix of an image I with p different pixel values with $n \times m$ pixels and a offset $(\delta x, \delta y)$. Then the values of $p \times p$ co-occurrence matrix C are given by

$$C_{(\delta x, \delta y)}(i, j) = \sum_{x=1}^{n} \sum_{y=1}^{m} \begin{cases} 1, & \text{if } I(x, y) = i \text{ and } I(x + \delta x, y + \delta y) = j \\ 0, & \text{otherwise} \end{cases}, \tag{5.24}$$

where (i, j) are the pixel values, (x, y) the spatial positions and $(x + \delta x, y + \delta y)$ are the spatial positions with the offsets. Thus $I(x, y)$ indicates the pixel value at (x, y) and $I(x + \delta x, y + \delta y)$ is the pixel value at $(x + \delta x, y + \delta y)$.

Note here that while (5.24) is for a greyscale image I, the same procedure can be applied to a colour image in each of its channels.

The code in listing 5.4 uses the command `mahotas.features.haralick` to compute the Haralick features (see figure 5.4 for the output).

```
# load packages
import mahotas
import mahotas.demos
import mahotas as mh
import numpy as np
from pylab import imshow , show

nuclear = mahotas.demos.nuclear_image ()
nuclear = nuclear [:,:,0]
nuclear = mahotas.gaussian_filter(nuclear ,90)
threshed = ( nuclear > nuclear.mean () )

labeled , n = mahotas.label(threshed)
imshow(labeled)
h_feature = mahotas.features.haralick(labeled)
imshow(h_feature)
show()
```

Listing 5.4 Haralick operation.

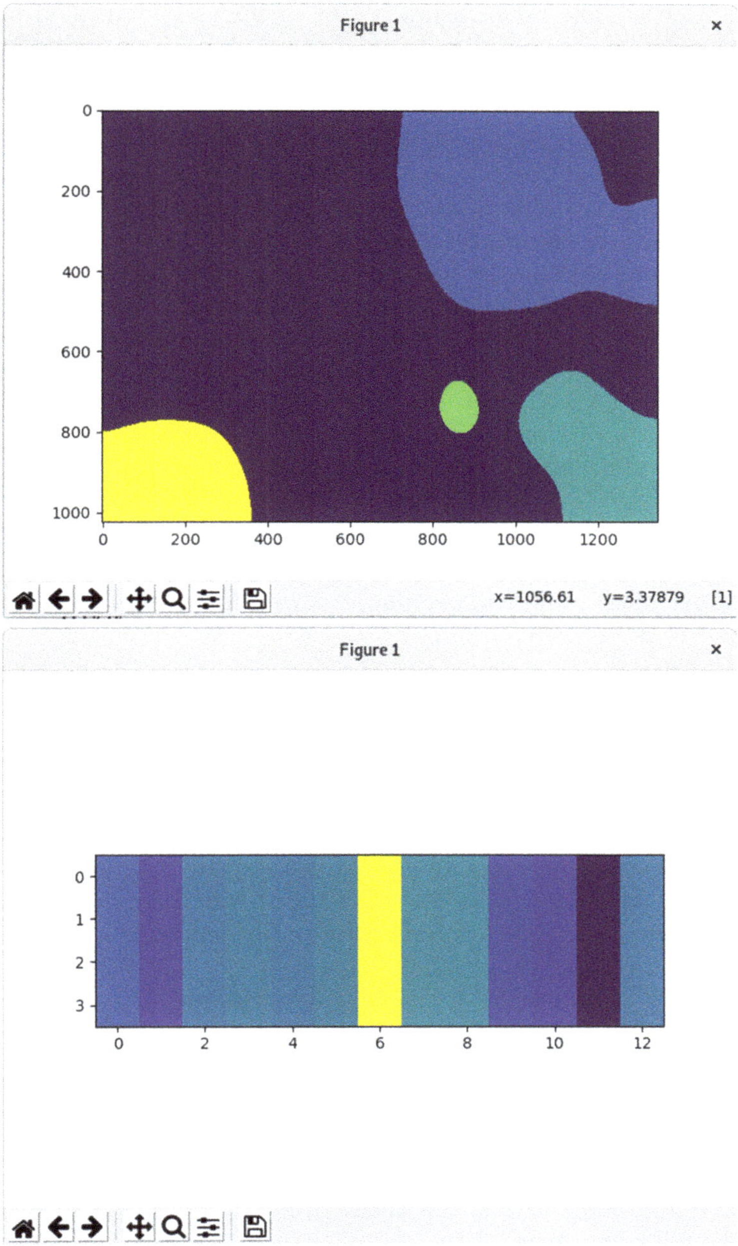

Figure 5.4. Output of listing 5.4: Haralick operation.

5.6 Local binary patterns

Local binary patterns are similar to Haralick features and are used to describe the texture of an image, among other applications. Local binary patterns are named such because they look for local patterns in the image and translate them into binary information. The binarization occurs in the neighbourhood of pixels by applying a threshold. The result is translated to a binary number placed on the centre pixel of the neighbourhood. Specifically, the steps of local binary patterns are the following:

1. Set a pixel value as centre pixel.
2. Set the neighbourhood pixels, for example, 3 × 3 matrix pixels in the neighbourhood. The total number of neighbourhood pixels is eight because the pixel at the centre is not counted.
3. Apply a threshold on the pixels of the neighbourhood. One if its values is greater than or equal to the centre pixel value, otherwise it is zero.
4. Then collect all of the threshold values of the neighbourhood pixels either clockwise or anti-clockwise. The collection is composed of ones or zeros; thus it is a binary code. Convert the binary code into decimal representation.
5. Replace the centre pixel with the decimal representation of the binary code.
6. Apply the same steps to all the pixels in the image.

The algorithm in `Python` of listing 5.5 computes the local binary patterns of an image for a 3 × 3 matrix as a neighbourhood (see figure 5.5 for the output).

5.7 Keypoint detectors

A keypoint and a descriptor define a feature of an image. A keypoint, or interest point, is delineated by some particular image intensity values around it, for example, a corner or an edge. A keypoint can be used for obtaining descriptors. However, it is essential to note that not every keypoint detector has a singular way of defining a descriptor. Recall that a descriptor is a finite vector which summarizes the properties of images, but a descriptor can also be employed for classifying the keypoint. Keypoints and descriptors can mutually define a feature of an image.

In this section we are going to describe some standard algorithms for detecting keypoints; they are also called keypoint detector algorithms.

5.7.1 FAST

The primary purpose of the FAST algorithm is to detect corners in images. The FAST algorithm was introduced in the paper 'Fusing points and lines for high-performance tracking' and later improved versions came with the publication of 'Machine learning for high-speed corner detection' (Rosten and Drummond 2006) and the 2008 paper 'Faster and better: a machine learning approach to corner detection' (Rosten *et al* 2008).

The FAST algorithm detects the edges over an image and to achive this it computes the following steps:

1. Select a pixel as centre pixel, the intensity value of which is p.
2. Draw a circle for which its centre is the centre pixel. The radius of the circle is usually taken as 3 px. Thus the number of pixels in the circumference is 16, as can be seen in figure 5.6.
3. Then, the code compares the intensity values of the circumference pixels and the centre pixel. If 3/4 of the pixels in the circumference are brighter than $p +$

```
import cv2 as ocv
import numpy as np
from matplotlib import pyplot as plt

def get_pixel(img, center, x, y):

    new_value = 0

    try:
        if img[x][y] >= center:
            new_value = 1

    except:
        pass

    return new_value

# Function for calculating LBP
def lbp_calculated_pixel(img, x, y):
    center = img[x][y]
    val_ar = []
    val_ar.append(get_pixel(img, center, x-1, y-1))
    val_ar.append(get_pixel(img, center, x-1, y))
    val_ar.append(get_pixel(img, center, x-1, y + 1))
    val_ar.append(get_pixel(img, center, x, y + 1))
    val_ar.append(get_pixel(img, center, x + 1, y + 1))
    val_ar.append(get_pixel(img, center, x + 1, y))
    val_ar.append(get_pixel(img, center, x + 1, y-1))
    val_ar.append(get_pixel(img, center, x, y-1))
    power_val = [1, 2, 4, 8, 16, 32, 64, 128]

    val = 0
    for i in range(len(val_ar)):
        val += val_ar[i] * power_val[i]

    return val


img_bgr = ocv.imread("image.jpg")

height, width, _ = img_bgr.shape
print(height)
print(width)

img_gray = ocv.cvtColor(img_bgr,
           ocv.COLOR_BGR2GRAY)

img_lbp = np.zeros((height, width),
                   np.uint8)
```

Listing 5.5 LBP operation.

```

for i in range(0, height):
    for j in range(0, width):
        img_lbp[i, j] = lbp_calculated_pixel(img_gray, i, j
          )

plt.imshow(img_lbp, cmap ="gray")
plt.show()
```

Listing 5.5 (Continued.)

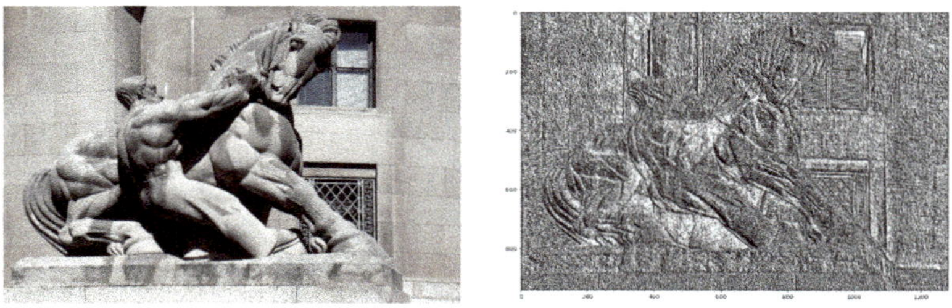

Original Image   Results of process

Figure 5.5. Output of listing 5.5: LBP operation. Credit: Highsmith, Carol M., 1946 - United States Library of Congress's Prints and Photographs Division under the digital ID highsm.09920.

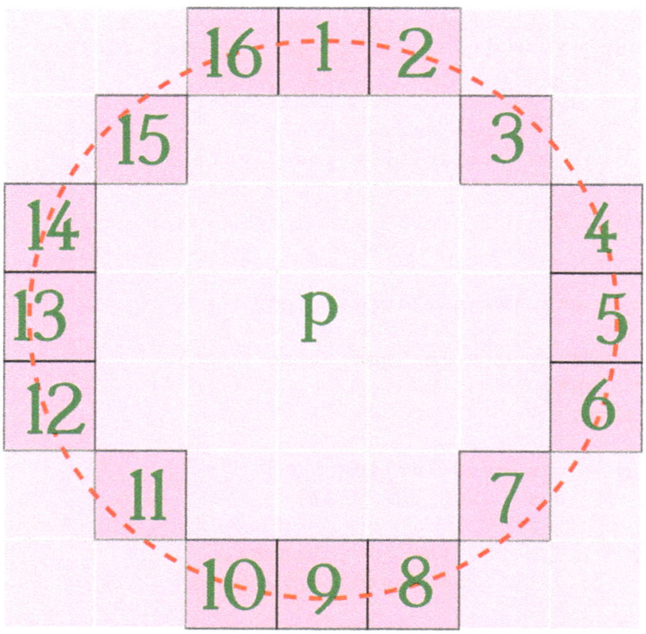

Figure 5.6. The FAST scheme.

t or darker than $p - t$, then this pixel is to be considered a keypoint. t is a threshold defined by the user.

4. Carry out the same process for all the pixels in the image to detect all the keypoints.

Note that this procedure is applied over a greyscale image. The `Python` code in lisiting 5.6 computes the keypoints with FAST (see figure 5.7 for the output).

Note that command `if` ensures that the code runs despite the possible different versions of `OpenCV`.

5.7.2 The Harris method

The next keypoint detector that we are going to study was presented in the paper 'A combined corner and edge detector' by Harris and Stephens (Harris and Stephens 1988).

```python
from __future__ import print_function
import numpy as np
import cv2 as ocv
import imutils

# load the image
img =ocv.imread("image.jpg")
ocv.imshow("Original",img)
# convert it to grayscale
gray =ocv.cvtColor(img,ocv.COLOR_BGR2GRAY)

# compute FAST keypoints in OpenCV 2.4
if imutils.is_cv2() :
    detector = ocv.FeatureDetector_create("FAST")
    kps = detector.detect(gray)

# otherwise , compute FAST keypoints in OpenCV 3+
else :
    detector = ocv.FastFeatureDetector_create()
    kps = detector.detect(gray,None)

print ( " # of keypoints : {} " . format ( len ( kps ) ) )

# draw all the keypoints
for kp in kps :
    r = int (0.5 * kp.size)
    (x,y)=np.int0(kp.pt)
    ocv.circle(img,(x,y),r,(127,0,127),1)

# show the keypoints
ocv.imshow( " Keypoints " , img )
ocv.waitKey()
```

Listing 5.6 The FAST algorithm.

Figure 5.7. Output of listing 5.6: FAST descriptor. Credit: Highsmith, Carol M., 1946 - United States Library of Congress's Prints and Photographs Division under the digital ID highsm.09920.

The Harris method detects whether the region is a keypoint or not with the gradients of the image and for that it uses the following matrix M:

$$M = \begin{bmatrix} \sum G_x^2 & \sum\sum G_x G_y \\ \sum\sum G_x G_y & \sum G_y^2 \end{bmatrix}. \tag{5.25}$$

Then the idea is to obtain the eigenvalue decomposition of the matrix. Depending on the eigenvalues λ_1 and λ_2 we will see if the region is a corner or not. The score R will tell us the nature of the region and is given by

$$R = \det(M) - k(\mathrm{trace}(M))^2, \tag{5.26}$$

where k is a constant and

$$\det(M) = \lambda_1\lambda_2, \quad \mathrm{trace}(M) = \lambda_1 + \lambda_2. \tag{5.27}$$

According to Harris, with the value of the score R we can conclude the following:

1. If $|R|$ is close to zero it means it is a flat region.
2. If R is negative there is an edge.
3. However, if λ_1 and λ_2 are quite similar then$|R|$ is large. Then the region we are talking about is a keypoint. See figure 5.8.

The code in listing 5.7 computes the Harris algorithm. Observe that the `if` ensures that the code runs despite the possible different versions of `OpenCV` (see figure 5.9 for the output).

5.7.3 GFTT

The Shi–Tomasi keypoint detector is firmly based on the Harris algorithm. Here we are going to call the Shi–Tomasi keypoint detector GFTT. GFTT is similar to the Harris algorithm, but it chooses a different score for R,

$$R = \min(\lambda_1, \lambda_2), \tag{5.28}$$

Figure 5.8. The HARRIS scheme.

where λ_1 and λ_2 are the eigenvalues of the matrix of equation (5.25). If the score R is greater than the user-defined threshold T then the region that we are examining is a corner.

GFTT detects the four following cases to distinguish if the region is a corner:

1. If λ_1 and λ_2 of a given region are less than T, then this region is not a keypoint.
2. If $\lambda_1 < T$ we do not have a keypoint.
3. If $\lambda_2 < T$ we also do not have a keypoint.
4. However, if $\lambda_1 < T$ and $\lambda_2 < T$ then we have a keypoint.

The code in listing 5.8 computes the GFTT keypoints over an image. Again note that the command `if` ensures that the code runs despite the possible different versions of `OpenCV` (see figure 5.10 for the output).

5.7.4 DoG

The difference of Gaussian (DoG) keypoint detector was elaborated by David Lowe and presented in his paper 'Object recognition from local scale-invariant features' in 1999 (Lowe 1999). The DoG algorithms use the following steps to obtain the keypoints of a given image:

1. The first step of the DoG keypoint detector is to produce fuzzy scale spatial images. Here we obtain the original image and generate progressively blurred variants of it. Then we split the dimensions of the image and repeat the whole process.

```
# load packages
from __future__ import print_function
import numpy as np
import cv2 as ocv
import imutils


def harris(gray,blockSize=2,apetureSize=3,k=0.1,T=0.02) :
    gray = np.float32(gray)
    H = ocv.cornerHarris(gray,blockSize,apetureSize,k)

    # keypoint size a 3 - pixel radius )
    kps = np.argwhere(H>T*H.max())
    kps = [ocv.KeyPoint(pt[1],pt[0],3) for pt in kps ]

    return kps


img = ocv.imread("image.jpg")
gray = ocv.cvtColor(img,ocv.COLOR_BGR2GRAY)

if imutils.is_cv2() :
    detector = ocv.FeatureDetector_create("HARRIS")
    kps = detector.detect( gray )

else :
    kps = harris ( gray )

print ( "␣#␣of␣keypoints␣:␣{}␣" . format ( len ( kps ) ) )


for kp in kps :
    r = int (0.5 * kp . size )
    (x,y) = np.int0( kp.pt )
    ocv.circle(img,(x,y) , r , (0, 127 , 255) , 5)

ocv.imshow( "␣Keypoints␣" , img )
ocv.waitKey()
```

Listing 5.7 The Harris descriptor.

Original Image　　Processed image

Figure 5.9. Output for listing 5.7: the Harris descriptor. Credit: Highsmith, Carol M., 1946 - United States Library of Congress's Prints and Photographs Division under the digital ID highsm.09920.

```
from __future__ import print_function
import numpy as np
import cv2 as ocv
import imutils

def gftt ( gray , maxCorners =0 , qualityLevel =0.01 ,
   minDistance =1 ,
            mask = None ,
              blockSize =3 ,
              useHarrisDetector = False ,
              k =0.04) :

    kps =ocv.goodFeaturesToTrack(gray,
                 maxCorners ,
                 qualityLevel ,
                 minDistance ,
                 mask = mask ,
                 blockSize = blockSize ,
     useHarrisDetector = useHarrisDetector , k = k )

    return
    [ocv.KeyPoint(pt[0][0],pt[0][1],3) for pt in kps ]

img =ocv.imread("image.jpg")
ocv.imshow("Original",img)
gray = ocv.cvtColor(img,ocv.COLOR_BGR2GRAY )

if imutils.is_cv2 () :
    detector = ocv.FeatureDetector_create( "␣GFTT␣" )
    kps = detector.detect ( gray )

else :
    kps = gftt ( gray )


for kp in kps :
    r = int (0.5 * kp . size )
    (x , y ) = np . int0 ( kp . pt )
    ocv.circle(img,(x, y),r,(255 , 255 ,0 ) , 3)

print ( "␣#␣of␣keypoints␣:␣{}␣" . format ( len ( kps ) ) )

ocv.imshow ( "␣keypoints␣" , img )
ocv.waitKey (0)
```

Listing 5.8 The GFTT descriptor.

2. Then we subtract two consecutive images of the same scale. Recall that we produce several images with the same scale with different blur. We carry this out for all the scales generated in the first step. This process can be seen in figure 5.11

Figure 5.10. Output for listing 5.8: the GFTT descriptor. Credit: Highsmith, Carol M., 1946 - United States Library of Congress's Prints and Photographs Division under the digital ID highsm.09920.

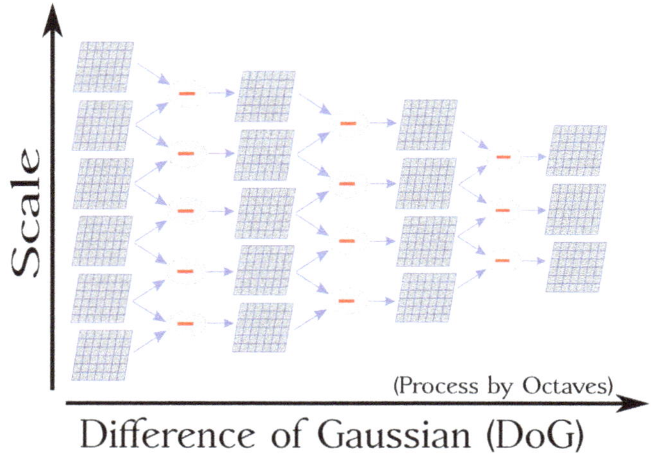

Figure 5.11. The DoG layers scheme.

3. Then we obtain the maxima and minima in the DoG images. The maxima or local minima are the largest/smallest value pixels in a defined neighbourhood. For each pair of DoG images we obtain local minima and maxima. The final step is to collect all the maxima and minima across all sizes of the images and label them as keypoints.

The code to compute the DoG keypoints in `Python` is shown in listing 5.9. Again observe the use of the command `if` to facilitate the two versions of calculating the DoG keypoints depending the different versions of `OpenCV` (see figure 5.12 for the output).

5.7.5 Fast Hessian

The next algorithm to study is fast Hessian. The fast Hessian algorithm is formulated on equivalent postulates to DoG. These postulates tell us that keypoints should be both repeatable and recognizable at various scales of an image, in other words, in the same image with several sizes. However, fast Hessian is quite distinct

```
from __future__ import print_function
import numpy as np
import cv2 as ocv
import imutils


img = ocv.imread("image.jpg")
gray = ocv.cvtColor(img,ocv.COLOR_BGR2GRAY )

if imutils.is_cv2() :
    detector =ocv.FeatureDetector_create( "␣SIFT␣" )
    kps = detector.detect ( gray )

else :
    detector = ocv.xfeatures2d.SIFT_create ()
    ( kps , _ ) = detector.detectAndCompute ( gray , None )

print ( "␣#␣of␣keypoints␣:␣{}␣" . format ( len ( kps ) ) )

for kp in kps :
    r = int (0.5 * kp . size )
    (x , y ) = np.int0 ( kp . pt )
    ocv.circle(img , (x , y ) , r , (0 , 255 , 255) , 2)

# show the image
ocv.imshow ( "␣keypoints␣" , img )
ocv.waitKey (0)
```

Listing 5.9 The DoG descriptor.

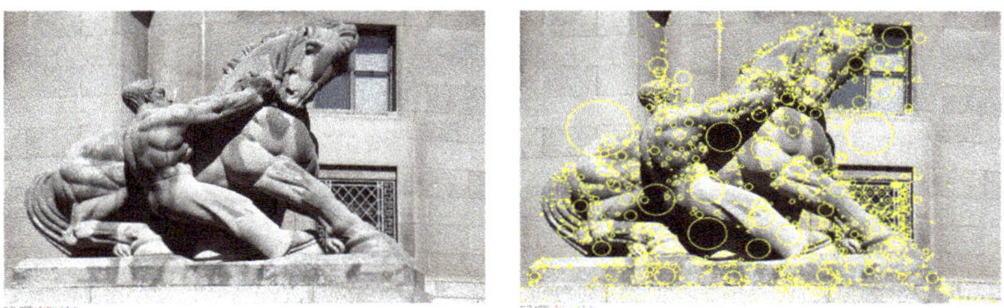

Original Image Processed image

Figure 5.12. Output for listing 5.9: the DoG descriptor. Credit: Highsmith, Carol M., 1946 - United States Library of Congress's Prints and Photographs Division under the digital ID highsm.09920.

from DoG, because in the first the difference is computed with Haar wavelets and in the second one difference is computed directly.

Despite the fact that the mathematical theory behind Haar wavelets is outside the scope of this chapter, we will provide some background. A Haar wavelet is a particular sequence of functions. This sequence was proposed in 1909 by Alfred Haar. Haar used these functions to give an example of an orthonormal accounting

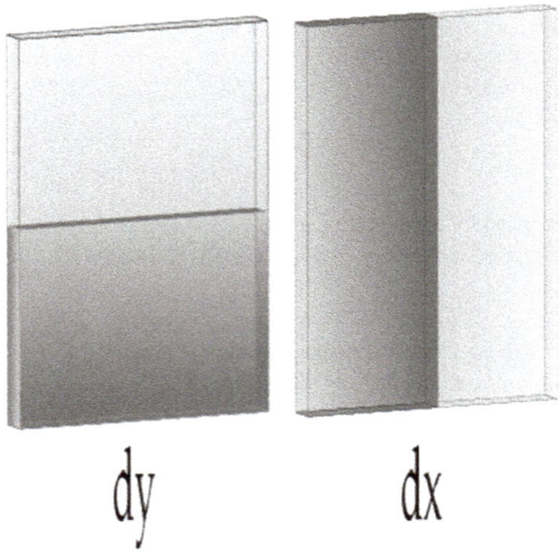

Figure 5.13. The Haar scheme.

system for the space of integrable square functions on the real line. Wavelet analysis is comparable to Fourier analysis in the way it provides an objective function over an interval to be represented in terms of an orthonormal basis. The same approach can be applied in artificial vision where the objective function is the image and we are interested in representing it in terms of an orthonormal basis. With this procedure we can detect the keypoints of the image. In this process the Hessian matrix is used, which is where the name of the algorithm comes from. See figure 5.13.

Thus to represent the image in terms of an orthonormal basis, we convolve the image with a Haar wavelet in several scales. Then the region to be denoted as a keypoint is the one for which the candidate score is higher than a $3 \times 3 \times 3$ neighbour, which covers its surrounding neighbours and the images of smaller and larger size.

Recall that our goal here is to briefly describe the algorithm and introduce the code in `Python`. The `Python` code to compute the fast Hessian keypoints is shown in listing 5.10. Once again, the command `if` serves to differentiate between the versions of `OpenCV` (see figure 5.14 for the output).

5.7.6 STAR

STAR is another algorithm for computing the keypoints over an image. The fundamentals of the STAR keypoint can be found in the 2008 paper 'CenSurE: center surround extremas for real-time feature detection and matching' (Agrawal *et al* 2008). Although STAR is not very popular, its performance is similar to the DoG but with increased speed. The difference between the two algorithms lies in that STAR uses octagon box filters and integral images in order to increase the speed of the computation. The STAR keypoint detector is implemented in `OpenCV`.

The code in listing 5.11 computes the keypoint of an image with STAR (see figure 5.15 for the output).

```python
from __future__ import print_function
import numpy as np
import cv2 as ocv
import imutils

img =ocv.imread ( "image.jpg" )
gray = ocv.cvtColor(img, ocv.COLOR_BGR2GRAY)

if imutils . is_cv2 () :
    detector =ocv.FeatureDetector_create( "␣SIFT␣" )
    kps = detector.detect ( gray )
else :
    detector = ocv.xfeatures2d.SIFT_create ()
    ( kps , _ ) = detector.detectAndCompute( gray , None )
print ( "␣#␣of␣keypoints␣:␣{}␣" . format ( len ( kps ) ) )

for kp in kps :
    r = int (0.5 * kp . size )
    (x , y ) = np . int0 ( kp . pt )
    ocv.circle ( img , (x , y ) , r , (125 ,55 , 255) , 2)

ocv.imshow ( "␣keypoints␣" , img )
ocv.waitKey (0)
```

Listing 5.10 The fast Hessian descriptor.

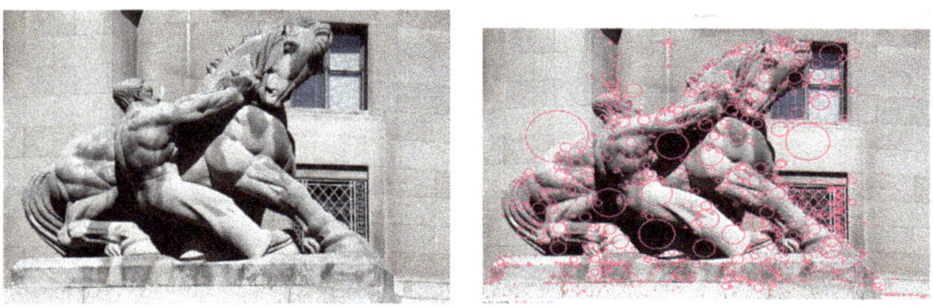

Original Image   Processed image

Figure 5.14. Output for listing 5.10: the fast Hessian descriptor. Credit: Highsmith, Carol M., 1946 - United States Library of Congress's Prints and Photographs Division under the digital ID highsm.09920.

5.7.7 MSER

The next algorithm we are going to study is maximally stable extremal regions (MSER). MSER computes several binarizations of a given image with different thresholds. The binarization is white to black and black to white. In the case of white to black, the first threshold values are such that the first resulting image is white. Then the threshold value is modified; each image turns out to have more black pixels. In the case of black to white the procedure is the the opposite of white to black.

Then MSER computes the following rules:

1. For all the resulting images, compute the connected component analysis. The connected components, in a 2D image, are clusters of pixels with the same value

```
from __future__ import print_function
import numpy as np
import cv2 as ocv
import imutils

img = ocv.imread ( "image.jpg" )
gray = ocv.cvtColor ( img ,ocv.COLOR_BGR2GRAY )

if imutils.is_cv2 () :
    detector = cv.FeatureDetector_create( "␣STAR␣" )
    kps = detector . detect ( gray )

else :
    detector = ocv.xfeatures2d.StarDetector_create ()
    kps = detector.detect ( gray )

print ( "␣#␣of␣keypoints␣:␣{}␣" . format ( len ( kps ) ) )

for kp in kps :
    r = int (0.5 * kp . size )
    (x , y ) = np . int0 ( kp . pt )
    ocv.circle ( img , (x , y ) , r , (255 , 127 ,0) , 2)

ocv.imshow ( "␣keypoints␣" , img )
ocv.waitKey (0)
```

Listing 5.11 The STAR descriptor.

Original Image Processed image

Figure 5.15. Output for listing 5.11: the STAR descriptor. Credit: Highsmith, Carol M., 1946 - United States Library of Congress's Prints and Photographs Division under the digital ID highsm.09920.

which are connected. This means that for white to black, the black connected components are labelled. For black to white, the white components are labelled.

2. Then MSER measures the areas of all the connected components over multiple threshold values.
3. The areas that remain approximately constant in size are considered to be keypoints.

It is important to note that MSER usually fails with distorted or blurred images. The code in listing 5.12 computes the MSER keypoints (see figure 5.16 for the output).

```
# load packages
from __future__ import print_function
import numpy as np
import cv2 as ocv
import imutils

# load the image
img = ocv.imread( "image.jpg" )
# convert it to grayscale
gray = ocv.cvtColor ( img ,ocv.COLOR_BGR2GRAY )

# compute MSER keypoints in OpenCV 2.4
if imutils.is_cv2 () :
    detector = ocv.FeatureDetector_create("MSER")
    kps = detector.detect ( gray )

# compute MSER keypoints in OpenCV 3+
else:
    detector = ocv.MSER_create()
    kps = detector.detect ( gray )

print ( "␣#␣of␣keypoints␣:␣{}␣" . format ( len ( kps ) ) )

# draw all keypoints
for kp in kps :
    r = int (0.5 * kp . size )
    (x , y ) = np . int0 ( kp . pt )
    ocv.circle ( img , (x , y ) , r , (0 , 255 , 127) , 2)

# show the image
ocv.imshow ( "␣keypoints␣" , img )
ocv.waitKey (0)
```

Listing 5.12 The MSER descriptor.

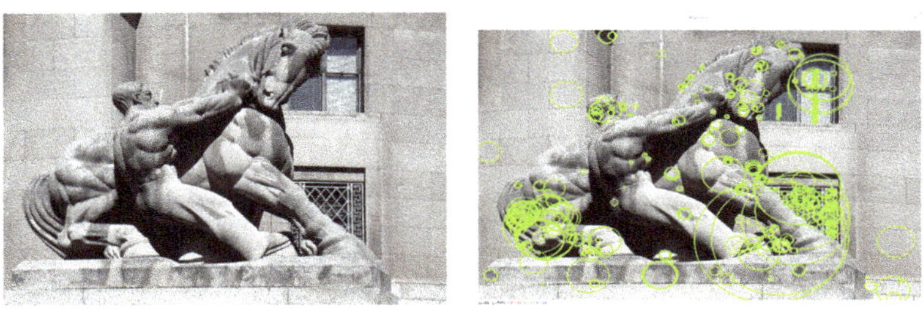

Original Image   Processed image

Figure 5.16. Output for listing 5.12: the MSER descriptor. Credit: Highsmith, Carol M., 1946 - United States Library of Congress's Prints and Photographs Division under the digital ID highsm.09920.

5.7.8 BRISK

The binary robust invariant scalable keypoint (BRISK) is a generalization of the algorithm FAST. The abstraction consists in that BRISK computes FAST over a pyramid image, while the original FAST is executed only in one image scale (Leutenegger *et al* 2011).

If a keypoint detected by FAST and prevails throughout the pyramid image, it is considered a keypoint by BRISK. The `Python` code for BRISK is shown in listing 5.13 (see figure 5.17 for the output).

5.7.9 ORB

The ORB keypoint detector, from the paper 'ORB: An efficient alternative to SIFT or SURF', is another extension to the FAST keypoint detector. ORB executes FAST over an image pyramid, like BRISK. Then the Harris keypoint score of section 5.7.2 is applied to rank and sort the keypoints. The final step of ORB is to identify the keypoints that are rotational invariants (Rublee *et al* 2011).

The code in listing 5.14 is the implementation of ORB in `Python` (see figure 5.18 for the output).

```
from __future__ import print_function
import numpy as np
import cv2 as ocv
import imutils

img = ocv.imread ( "image.jpg" )
gray =ocv.cvtColor( img ,ocv.COLOR_BGR2GRAY )

# compute BRISK keypoints in OpenCV 2.4
if imutils.is_cv2() :
    detector =ocv.FeatureDetector_create("BRISK")
    kps = detector . detect ( gray )

# compute BRISK keypoints in OpenCV 3+
else :
    detector = ocv.BRISK_create ()
    kps = detector.detect ( gray )

print ( "␣#␣of␣keypoints␣:␣{}␣" . format ( len ( kps ) ) )

# draw all keypoints
for kp in kps :
    r = int (0.5 * kp . size )
    (x , y ) = np . int0 ( kp . pt )
    ocv.circle ( img , (x , y ) , r , (127 ,255 ,255) , 2)

ocv.imshow ( "␣keypoints␣" , img )
ocv.waitKey (0)
```

Listing 5.13 The BRISK descriptor.

Figure 5.17. Output for listing 5.13: the BRISK descriptor. Credit: Highsmith, Carol M., 1946 - United States Library of Congress's Prints and Photographs Division under the digital ID highsm.09920.

```
from __future__ import print_function
import numpy as np
import cv2 as ocv
import imutils

img = ocv.imread ( "image.jpg" )
gray = ocv.cvtColor ( img ,ocv.COLOR_BGR2GRAY )

if imutils.is_cv2 () :
    detector = ocv.FeatureDetector_create("␣ORB␣")
    kps = detector.detect ( gray )
else :
    detector = ocv. ORB_create ()
    kps = detector.detect ( gray )

print ( "␣#␣of␣keypoints␣:␣{}␣" . format ( len ( kps ) ) )

for kp in kps :
    r = int (0.5 * kp . size )
    (x , y ) = np . int0 ( kp . pt )
    ocv.circle ( img , (x , y ) , r , (127 , 255 , 127) ,
       2)

ocv.imshow ( "␣keypoints␣" , img )
ocv.waitKey (0)
```

Listing 5.14 The ORB descriptor.

Figure 5.18. Output for listing 5.14: the ORB descriptor. Credit: Highsmith, Carol M., 1946 - United States Library of Congress's Prints and Photographs Division under the digital ID highsm.09920.

5.8 Local invariant descriptors

Once keypoints have been detected by the algorithms of the previous section, it is time to describe and quantify the region of the image around the detected keypoint. The algorithms in this section achieve this task.

5.8.1 SIFT

The SIFT descriptor has as its input the image and the set of input keypoints. It takes a region of 16×16 pixels around any keypoint. SIFT divides each region of 16×16 pixels into 4×4 pixel windows.

For each 4×4 region, the gradient and the direction are computed with equations (2.8) and (2.9), respectively. Then a gradient oriented histogram is computed for each of the sixteen 4×4 regions. The final step is to concatenate the 16 gradient oriented histograms and form the feature vector.

The code in listing 5.15 computes the SIFT descriptor (see figure 5.19 for the output).

5.8.2 SURF

Now let us focus on the speeded up robust features (SURF) descriptor. SURF is a high-performance detector and descriptor of the points of interest in an image, where

```
from __future__ import print_function
import numpy as np
import cv2 as ocv
import imutils

image = ocv.imread ( "image.jpg" )
img= ocv.cvtColor( image ,ocv.COLOR_BGR2GRAY )

if imutils.is_cv2 () :
    detector = ocv.FeatureDetector_create("SIFT")
    extractor = ocv.DescriptorExtractor_create("SIFT")
    kps = detector.detect(img)
    ( kps , descs ) = extractor.compute ( img , kps )

else :
    detector = ocv.xfeatures2d.SIFT_create ()
    ( kps , descs ) = detector.detectAndCompute(img,None)

for kp in kps :
    r = int (0.5 * kp . size )
    (x , y ) = np . int0 ( kp . pt )
    ocv.circle ( img , (x , y ) , r , (0 ,255, 255) , 1)

ocv.imshow ( "SIFT" , img )
ocv.waitKey (0)
```

Listing 5.15 The SIFT descriptor.

Original Image   Processed image

Figure 5.19. Output for listing 5.15: the SIFT descriptor. Credit: Highsmith, Carol M., 1946 - United States Library of Congress's Prints and Photographs Division under the digital ID highsm.09920.

the image is transformed into coordinates using a technique called multi-resolution (Bay *et al* 2006). It consists of making a replica of the original image in a Gaussian pyramidal or Laplacian pyramidal shape, and obtaining images of the same size but with reduced bandwidth. In this way a blurring effect is achieved on the original image, called scale-space. This technique ensures that the points of interest are invariant in the scaling. The SURF algorithm is based on its predecessor SIFT. See figure 5.13.

The SURF and SIFT descriptors have the same first steps. They take a region of 16×16 pixels around any keypoint. Each region of 16×16 pixels is divided into 4×4 pixel windows. Thereafter the SURF and SIFT descriptors diverge.

SURF applies a Haar wavelet response on each 4×4 pixel window. The Haar wavelet response is computed by SURF in the horizontal and vertical directions. dx is the Haar wavelet response over the horizontal direction and dy is the Haar wavelet response over the vertical direction. Then SURF weights the response with a Gaussian kernel centred at the point of interest. See figure 4.3.

The final step is that SURF computes the following feature vector for each 4×4 pixel window:

$$v = \left[\sum dx,\ \sum dy,\ \sum |dx|,\ \sum |dy|\right], \tag{5.29}$$

where $\sum dx$ is the sum over the values of dx and $\sum dx$ is the sum of values of dy. $|dx|$ and $|dy|$ are the absolute values of dx and dy, respectively.

The code in listing 5.16 executes the SURF descriptor (see figure 5.20 for the output).

5.9 Binary descriptors

The last type of descriptor that we are going to study is the so-called binary descriptors. They are called such because the feature vectors consist of strings of zeros and ones. A significant advantage of binary descriptors is that they are fast to compute so they are ideal for real-time applications, since computing binary operations is more rapid than calculating the Euclidean distance for 128 dim.

```python
from __future__ import print_function
import numpy as np
import cv2 as ocv
import imutils

image = ocv.imread ( "image.jpg" )
img= ocv.cvtColor ( image ,ocv.COLOR_BGR2GRAY )

if imutils.is_cv2() :
    detector = ocv.FeatureDetector_create("SURF")
    extractor = ocv.DescriptorExtractor_create("SURF")

    kps = detector.detect (img )
    ( kps , descs ) = extractor.compute ( img , kps )

else :
    detector = ocv.xfeatures2d.SURF_create()
    ( kps , descs ) =
     detector.detectAndCompute(img , None )

for kp in kps :
    r = int (0.5 * kp . size )
    (x , y ) = np . int0 ( kp . pt )
    ocv.circle ( img , (x , y ) , r , (127 , 255 , 127) ,
      2)

ocv.imshow ( " keypoints " , img )
ocv.waitKey (0)
```

Listing 5.16 The SURF descriptor.

Original Image　　Processed image

Figure 5.20. Output for listing 5.16: the SURF descriptor. Credit: Highsmith, Carol M., 1946 - United States Library of Congress's Prints and Photographs Division under the digital ID highsm.09920.

The different binary descriptors share almost the same strategy:

1. All of them use a sampling pattern to extract pixels around the keypoint. The properties of the sampling pattern deeply influence the performance of binary descriptor algorithms.

2. Once the pixel is collected the binary descriptor algorithms should compensate for rotation. This means that no matter how rotated the area around the keypoint is, the resulting feature vector should be the same.
3. From the sampling pattern extraction, two sampling pairs are compared. A sampling pair is two sets of pixels taken from different sampling patterns around the keypoint.

The comparison between the two pixels of a sampling pair is very simple and fast. Given the intensity values of these two pixels a_1 and a_2, the following comparison takes place:

$$g(a_1, a_2) = \begin{cases} a_1 >= a_2 \Rightarrow 1 \\ a_1 < a_2 \Rightarrow 0 \end{cases}, \tag{5.30}$$

where $g(a_1, a_2)$ is called the comparison function. In the following subsection we are going to study the primary binary descriptors that follow this approach.

5.9.1 BRIEF

The first binary descriptor that we are going to study is binary robust independent elementary features (BRIEF). We mentioned earlier that the sampling pattern profoundly influences the performance of binary descriptors. This means that for different sampling patterns we may need another binary descriptor. However, BRIEF does not have a defined sampling pattern, instead it uses random sampling and extracts pixels around the keypoint. BRIEF uses a Gaussian distribution for the sampling, after applying a Gaussian blur filter over the image.

Once BRIEF obtains the sampling a comparison is performed with equation (5.30). This final step generates the binary feature vector. The code in listing 5.17 computes the BRIEF binary descriptor in `Python` (see figure 5.21 for the ouput) (Calonder *et al* 2010).

5.9.2 ORB binary descriptor

The ORB binary descriptor takes a keypoint and its surrounding area and computes the centre of mass of the region,

$$C = [M_{10}/M_{00}, M_{01}/M_{00}], \tag{5.31}$$

where $M_{i,j}$ are the moments of the surrounding area, that we previously studied with equation (5.1),

$$M_{i,j} = \sum_x \sum_y x^i y^j I(x, y). \tag{5.1}$$

With these moments ORB computes the orientation θ:

$$\theta = \arctan(M_{01}/M_{10}). \tag{5.32}$$

```
#load packages
from __future__ import print_function
import numpy as np
import cv2 as ocv
import imutils

#load image and convert gray
image = ocv.imread ( "image.jpg" )
img = ocv.cvtColor ( image , ocv.COLOR_BGR2GRAY )

# compute BRIEF keypoints in OpenCV 2.4
if imutils . is_cv2 () :
    detector = ocv.FeatureDetector_create("FAST" )
    extractor = ocv.DescriptorExtractor_create("BRIEF" )

# compute BRIEF keypoints in OpenCV 3+
else :
    detector = ocv.FastFeatureDetector_create()
    extractor = ocv.xfeatures2d.
       BriefDescriptorExtractor_create()

    # detect descriptors
    kps = detector . detect ( img )
    # extract descriptors
    ( kps , descs ) = extractor.compute ( img , kps )

# print keypoints
print ( "␣keypoints␣" . format ( len ( kps ) ) )
# print feature vector
print ( "␣feature␣vector␣" . format ( descs . shape ) )


# draw all keypoints
for kp in kps :
    r = int (0.5 * kp . size )
    (x , y ) = np . int0 ( kp . pt )
    ocv.circle ( img , (x , y ) , r , (0, 150 ,250) , 2)

ocv.imshow ( "␣keypoints␣" , img )
ocv.waitKey (0)
```

Listing 5.17 The BRIEF descriptor.

With the orientation θ we can rotate the surrounding area such that it is aligned with the x-axis. Then ORB executes steps similar to BRIEF to obtain the feature vector with the comparison function $g(a_1, a_2)$, equation (5.30). The code in listing 5.18 executes the ORB binary descriptors (see figure 5.22 for the ouput).

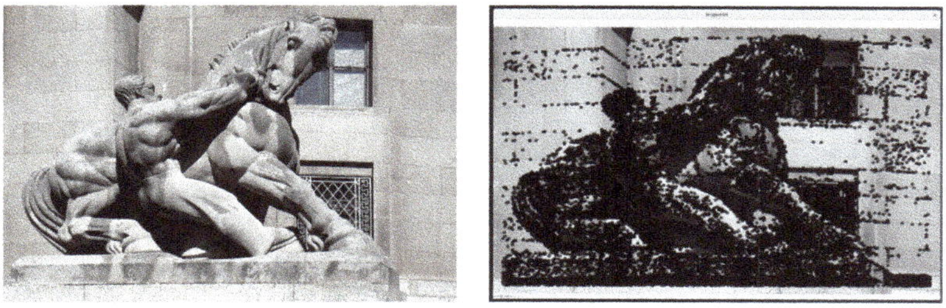

Original Image Processed image

Figure 5.21. Output for listing 5.17: the BRIEF descriptor. Credit: Highsmith, Carol M., 1946 - United States Library of Congress's Prints and Photographs Division under the digital ID highsm.09920.

```
from __future__ import print_function
import numpy as np
import cv2 as ocv
import imutils

image = ocv.imread ( "image.jpg" )
img = ocv.cvtColor ( image , ocv.COLOR_BGR2GRAY )

if imutils . is_cv2 () :
    detector = ocv.FeatureDetector_create("ORB")
    extractor = ocv.DescriptorExtractor_create("ORB")

    kps = detector . detect ( img )
    ( kps , descs ) = extractor . compute ( img , kps )

else :
    detector = ocv.ORB_create ()
    ( kps , descs ) = detector.detectAndCompute ( img ,
       None )

for kp in kps :
    r = int (0.5 * kp . size )
    (x , y ) = np . int0 ( kp . pt )
    ocv.circle ( img , (x , y ) , r , (127 , 255 , 127) ,
       2)

ocv.imshow ( " keypoints " , img )
ocv.waitKey (0)
```

Listing 5.18 The ORB binary descriptor.

5.9.3 The BRISK binary descriptor

The next descriptor is the BRISK binary descriptor. BRISK differs from BRIEF in its sampling pattern. While BRIEF randomly samples the pixels, BRISK uses a sampling pattern based on concentric rings. See figure 5.23.

BRISK uses long pairs and short pairs. Long pairs are used to compute/estimate the orientation of the pixel. Short pairs are compared to build the binary feature vector.

Figure 5.22. Output for listing 5.18: the SIFT descriptor. Credit: Highsmith, Carol M., 1946 - United States Library of Congress's Prints and Photographs Division under the digital ID highsm.09920.

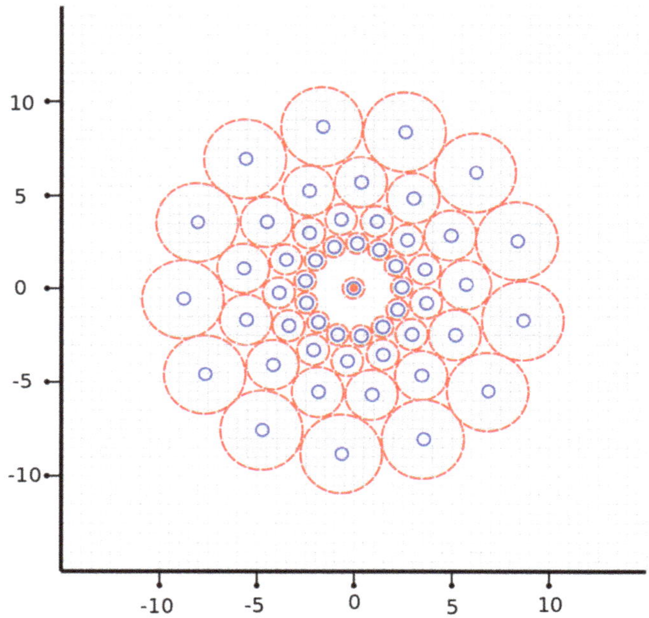

Figure 5.23. The BRISK binary descriptor scheme.

The Euclidean distance is computed for each pair to detect if it is a long pair or short pair with respect to the threshold. Once we know which ones the long pairs are, we use them to obtain the local gradients G_x and G_y with equations (2.6) and (2.7). The next step is to compute the orientation

$$\Theta = \arctan\left(\frac{G_x}{G_y}\right). \tag{5.33}$$

With Θ we can align the surrounding area to the x-axis and with the short pairs obtain the feature vector with the comparison function $g(a_1, a_2)$, equation (5.30).

```
from __future__ import print_function
import numpy as np
import cv2 as ocv
import imutils

image = ocv.imread ( "image.jpg" )
img = ocv.cvtColor(image , ocv.COLOR_BGR2GRAY )

if imutils.is_cv2 () :
    detector = ocv.FeatureDetector_create("BRISK")
    extractor = ocv.DescriptorExtractor_create("BRISK")

    kps = detector . detect (img )
    ( kps , descs ) = extractor . compute (img , kps )

else:
    detector = ocv.BRISK_create ()
    ( kps , descs ) = detector.detectAndCompute ( img , None )

# print keypoints
print("␣keypoints␣".format(len(kps)))
# print feature vector
print ( "␣feature␣vector␣" . format ( descs . shape ) )

for kp in kps :
    r = int (0.5 * kp . size )
    (x , y ) = np . int0 ( kp . pt )
    ocv.circle ( img , (x , y ) , r , (50 ,0 , 127) , 2)

ocv.imshow ( "␣keypoints␣" , img )
ocv.waitKey (0)
```

Listing 5.19 The BRIEF descriptor.

The code in listing 5.19 is the BRISK binary descriptor in `Python` (see figure 5.24 for the output).

5.9.4 FREAK

FREAK is the last algorithm that we are going to study in this chapter. FREAK is very similar to BRISK, but they differ in the sampling pattern. The human eye inspired the sampling pattern of FREAK. The sampling pattern can be seen in figure 5.25, where we have overlapping of circular regions. The number of circles per unit area is higher at the centre of the sampling pattern (Alahi *et al* 2012).

Using the described sampling pattern, FREAK obtains a coarse-to-fine approach very similar to the ORB binary descriptor. FREAK also compensates for the orientation, similar to BRISK, but with 45 predefined symmetric sampling patterns.

Figure 5.24. Output for listing 5.19: the BRIEF descriptor. Credit: Highsmith, Carol M., 1946 - United States Library of Congress's Prints and Photographs Division under the digital ID highsm.09920.

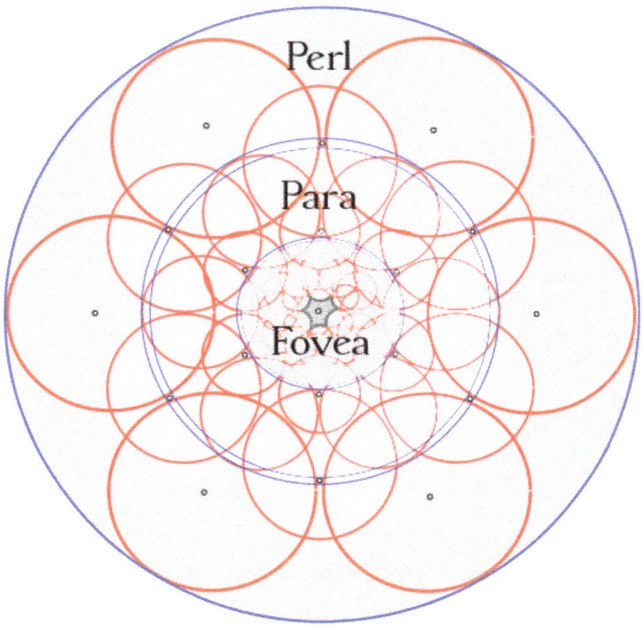

Figure 5.25. The FRISK binary descriptor scheme.

FREAK uses the intensity of the pixel to obtain the binary string, like BRISK and ORB.

The code in listing 5.20 executes the FREAK descriptor algorithm (see figure 5.26 for the output).

5.10 End notes

In this chapter we focus on studying image descriptors. For all of them we present the relevant `Python` code. We started with basic image statistics and then move on

```
# load packages
from __future__ import print_function
import numpy as np
import cv2 as ocv
import imutils

# load the image
image = ocv.imread ( "image.jpg" )
# convert it to grayscale
img = ocv.cvtColor(image , ocv.COLOR_BGR2GRAY )

# compute FREAK keypoints in OpenCV 2.4
if imutils.is_cv2 () :
    detector = ocv.FeatureDetector_create("FAST")
    extractor = ocv.DescriptorExtractor_create("FREAK")

    kps = detector.detect (img)
    (kps,descs)= extractor.compute (img,kps)

# compute FREAK keypoints in OpenCV 3+
else :
    detector = ocv.FastFeatureDetector_create()
    extractor = ocv.xfeatures2d.FREAK_create ()

    kps = detector.detect (img , None )
    ( kps , descs ) = extractor.compute (img, kps )

# print keypoints
print ( "␣keypoints␣".format ( len ( kps ) ) )
# print feature vector
print ( "␣feature␣vector".format ( descs . shape ) )


# draw all keypoints
for kp in kps :
    r = int (0.5 * kp . size )
    (x , y ) = np . int0 ( kp . pt )
    ocv.circle ( img , (x , y ) , r , (50 ,0 , 127) , 2)
# show the image
ocv.imshow ( "␣keypoints␣" , img )
ocv.waitKey (0)
```

Listing 5.20 The FREAK descriptor.

to robust image descriptors and keypoint detectors such as GFTT, FAST, DoG, fast Hessian, ORB, FREAK, START, etc.

Through the study of all these codes, we have shown their paradigms. We also saw how useful the keypoints are.

Original Image Processed image

Figure 5.26. Output of listing 5.20: the FREAK descriptor. Credit: Highsmith, Carol M., 1946 - United States Library of Congress's Prints and Photographs Division under the digital ID highsm.09920.

References

Agrawal M, Konolige K and Blas M R 2008 Censure: center surround extremas for realtime feature detection and matching *European Conference on Computer Vision* (Berlin: Springer) pp 102–15

Alahi A, Ortiz R and Vandergheynst P 2012 FREAK: fast retina keypoint *2012 IEEE Conf. on Computer Vision and Pattern Recognition* (Piscataway, NJ: IEEE) pp 510–7

Bay H, Tuytelaars T and Van Gool L 2006 SURF: speeded up robust features *European Conf. on Computer Vision* (Berlin: Springer) pp 404–17

Calonder M, Lepetit V, Strecha C and Fua P 2010 BRIEF: binary robust independent elementary features *European Conf. on Computer Vision* (Berlin: Springer) pp 778–92

Chityala R and Pudipeddi S 2020 *Image Processing and Acquisition Using* `Python` (Boca Raton, FL: CRC Press)

Forsyth D A and Ponce J 2002 *Computer Vision: A Modern Approach* (Englewood Cliffs, NJ: Prentice Hall)

Gonzalez R C, Woods R E and Eddins S L 2004 *Digital Image Processing Using MATLAB* (New Delhi: Pearson Education India)

Granlund G H and Knutsson H 2013 *Signal Processing for Computer Vision* (Berlin: Springer)

Haralick R M, Shanmugam K and Dinstein I H 1973 Textural features for image classification *IEEE Trans. Syst. Man Cybern.* **6** 610–21

Harris C G *et al* 1988 A combined corner and edge detector *Alvey Vision Conference* vol 15 pp 10–5244

Hu M-K 1962 Visual pattern recognition by moment invariants *IRE Trans. Inform. Theory* **8** 179–87

Klette R 2014 *Concise Computer Vision* (Berlin: Springer)

Leutenegger S, Chli M and Siegwart R Y 2011 BRISK: Binary robust invariant scalable keypoints *2011 International Conf. on Computer Vision* (Piscataway, NJ: IEEE) pp 2548–55

Lowe D G 1999 Object recognition from local scale-invariant features *Proc. of the Seventh IEEE International Conf. on Computer Vision* (Piscataway, NJ: IEEE) pp 1150–7

Lutz M 2001 *Programming* `Python` (Sebastopol, CA: O'Reilly)

Rosten E and Drummond T 2006 Machine learning for high-speed corner detection *European Conf. on Computer Vision* (Berlin: Springer)

Rosten E, Porter R and Drummond T 2008 Faster and better: a machine learning approach to corner detection *IEEE Trans. Pattern Anal. Mach. Intell.* **32** 105–19

Rublee E, Rabaud V, Konolige K and Bradski G 2011 ORB: An efficient alternative to SIFT or SURF *2011 Int. Conf. on Computer Vision* (Piscataway, NJ: IEEE) pp 2564–71
Schalkoff R J 1989 *Digital Image Processing and Computer Vision* vol 286 (New York: Wiley)
Szeliski R 2010 *Computer Vision: Algorithms and Applications* (Berlin: Springer)
Teague M R 1980 Image analysis via the general theory of moments *JOSA* **70** 920–30
Umesh P 2012 Image processing in `Python` *CSI Communications* **23** 2
Van Rossum G and Drake F L Jr 1995 `Python` *Tutorial* vol 620 (Amsterdam: Centrum voor Wiskunde en Informatica)

Rafael G González-Acuña, Héctor A Chaparro-Romo and Israel Melendez-Montoya

Chapter 6

Neural networks

Neural networks have been around for quite some time, the logic that neurons follow was constructed into theorems under several assumptions. Since then, a variety of neural networks have been constructed, many with specific purposes and all of them with *learning* capabilities. In recent years, due to hardware optimization in tensor product calculations and novel neural network architectures, the viability of using neural networks for various artificial vision and non-artificial vision tasks (Kim 2014, Pope *et al* 2019) using in real-time and in-cloud inference have become a reality and their applications have been growing. The theory and implementation behind the methods that are used in artificial vision are discussed in this chapter.

6.1 Introduction

The main idea behind the 1943 McCulloh–Pitts theory of neural activity (Pitts and McCulloch 1943) is that neurons are of an *all-or-nothing* character, which means that neurons are treated as binary threshold operators for different inputs. This did not provide great variety for the types of problems that this neural net logic could solve. In the McCulloh–Pitts logical model of neuron activity they proposed what we know today as an *architecture*, which contains many possibilities of how different *nets* of neurons may have inputs, outputs and are connected among themselves.

It was not until later in 1957 that psychologist Frank Rosenblatt (Rosenblatt 1958) published a report that contained the theory and electronic construction of a human-like thinking device. Rosenblatt returned to the McCulloh–Pitts neuronal activity model but introduced a continuous approach rather than a Boolean one. Rosenblatt contributed and developed much of this '*threshold logic*' and implemented a so-called *perceptron* (which is an analogue of a human neuron) that had *learning* properties and could linearly classify objects in an IBM machine at the Cornell Aeronautical Laboratory.

doi:10.1088/978-0-7503-3707-6ch6

The limitations of these single layer *perceptrons* were later mentioned in the book *Perceptrons* in 1969 (Minsky and Papert 1969) and later again in 1988 (Minsky and Papert 1988) by Minksy and Papert, where the necessity of a mathematical description of any image to be processed by a perceptron layer was discussed. The mathematical link that analytically describes images could not be generalized to (Yehudai and Shamir 2020), for example, pictures of cats and dogs and was later substituted by the generalization of features obtained based on multiple training examples without needing to explicitly describe them (Mamalet and Garcia 2012). In a sense, this is why deep learning algorithms are sometimes seen as a magical black box where thousands of data inputs can intrinsically make the classification process.

6.2 Neural networks in a nutshell

Let us dive directly into the perceptron model for classification. Let us say that we are looking to approximate a function f_* with an n-dimensional input vector

$$\mathbf{x} = (x_1\ x_2\ x_3\ \dots\ x_n) \tag{6.1}$$

so that a classification is given as y in the form

$$y = f_*(\mathbf{x}), \tag{6.2}$$

where y should take on discrete values. The main idea presented by Rosenblatt for a perceptron to learn a linear classification was to have inputs represented as a linearly independent vector $\mathbf{x}$ where the coefficients are called *weights*, in this case $\mathbf{W} = (w_1\ w_2\ w_3\ \dots w_n)$ and there exists a threshold value θ such that

$$f(x_1w_1 + x_2w_2 + x_3w_3 + \cdots + x_nw_n) = f(\mathbf{w}\cdot\mathbf{x}) = \begin{cases} 1, \text{ if } \mathbf{w}\cdot\mathbf{x} \geqslant \theta \\ 0, \text{ if } \mathbf{w}\cdot\mathbf{x} < \theta \end{cases}. \tag{6.3}$$

If we use a Heaviside activation function such as that in equation (6.3), then we obtain a classification threshold *hyperplane* of the form

$$x_1w_1 + x_2w_2 + x_3w_3 + \cdots + x_nw_n = \theta, \tag{6.4}$$

where this plane can linearly separate two classes (0 and 1) in N dimensions. We visualize two of these cases in figure 6.1.

The perceptron may be visualized as figure 6.2 and may be connected in series and in parallel to functions as weighted logical gates. In this manner we would obtain

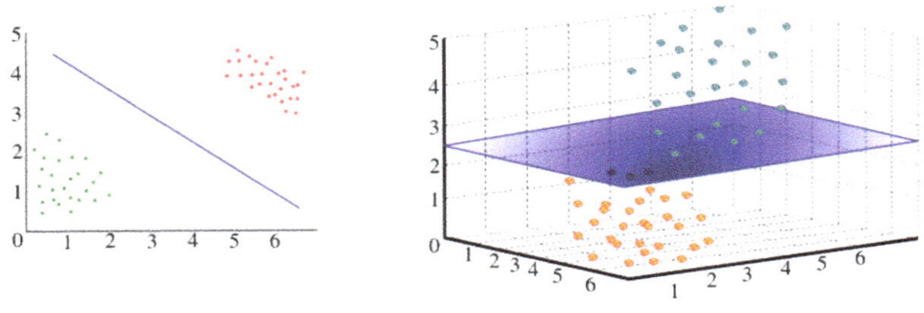

Figure 6.1. $\mathcal{R}^2$ (left) and $\mathcal{R}^3$ (right) example hyperplanes.

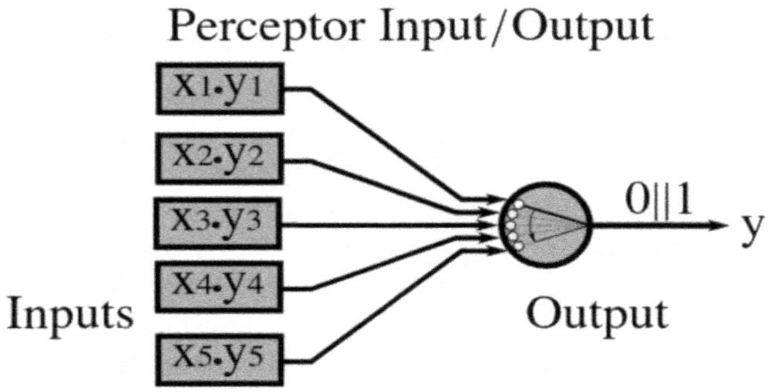

Figure 6.2. A visualization of the perceptron model.

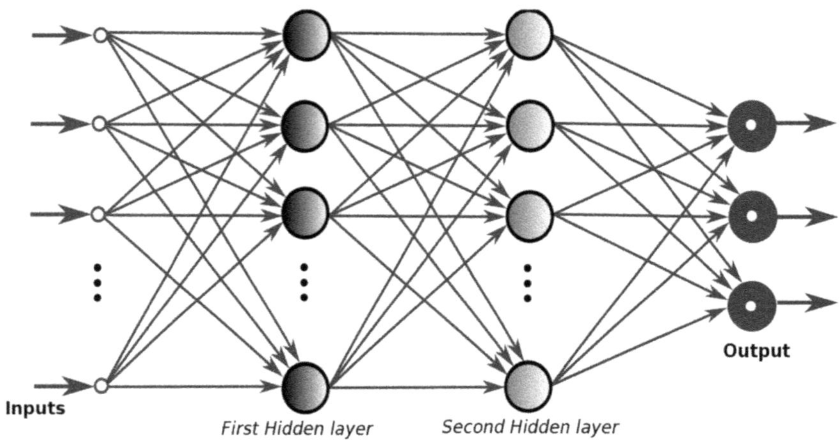

Figure 6.3. A visualization of a typical deep artificial neural network.

multiple layers of perceptrons, as can be seen in figure 6.3, with the following expression:

$$\mathbf{y} = f(\mathbf{x}) = f_1(f_2(f_3(\dots f_n(\mathbf{w} \cdot \mathbf{x})))), \tag{6.5}$$

where **y** is the output vector. It is said that as the number of composite functions increases then the neural network is *deeper*, and each function that composes $f(\mathbf{x})$ is called a *layer* of the neural network. By convention, the input **x** is the *first layer*, the last layer is called the *output layer* and the rest of the layers are called the *hidden layers.*

To extend this linear input model **x** to introduce non-linearity we may use a non-linear transformation such that the input is now a non-linear $\phi(\mathbf{x})$. Examples of non-linear transformations include $\sqrt{\mathbf{x}}$ and $\mathbf{x} \cdot \mathbf{xx}$, these non-linear transformations are known as *activation functions* (Glorot *et al* 2011).

A strategy that deep learning implies is that having information about the variables $y = f(\mathbf{x}; \theta, \mathbf{w})$ we may use a function ϕ such that $y = \phi(\mathbf{x}; \theta)\mathbf{w}$ may approximate f_* sufficiently.

6.3 Single perceptron learning

The main idea behind supervised classification training in machine learning is to be able to correctly classify a set of test data using certain training data that may statistically allow us to make conclusions. This can be achieved by optimizing certain parameters using classification metrics for final evaluations of the model. The test dataset is data that the system has never seen and the system will be prompted to make inferences using statistical similarities in the data classes from the training dataset. A definition of machine learning given by Tom M Mitchell is as follows: '*A computer program is said to learn from experience E with respect to some class of tasks T and performance measure P if its performance at tasks T, as measured by P, improves with experience E* (Mitchell *et al* 1990).

Binary perceptron

Let us use a simple linear classification task to explain the basics of training a classification perceptron. Let us suppose we want to find the hyperplane that linearly separates two groups of linearly separable points A and B contained in a matrix $\mathbf{x} = (\mathbf{x}_1, \mathbf{x}_2, \mathbf{x}_3, \dots, \mathbf{x}_N) \in \mathcal{R}^2$ on a 2D Cartesian plane and every vector $(\mathbf{x}_1, \mathbf{x}_2, \mathbf{x}_3, \dots, \mathbf{x}_N)$. So then the total dataset contains N points and they belong to either a class -1 or $+1$ represented by $\mathbf{y} = (y_1, y_2, y_3, \dots, y_n)$. A single perceptron model can perform this classification task if we define a function

$$f(\mathbf{x}, \mathbf{w}) = \mathbf{w} \cdot \mathbf{x} = z, \tag{6.6}$$

where $\mathbf{x}$ is an $2 \times N$ input vector, $\mathbf{w}$ is a $N \times 2$ weighted coefficients vector and $y(z)$, our binary classification function, will be defined as

$$y(z) = \begin{cases} +1, \text{if } z \geqslant \theta \\ -1, \text{if } z < \theta \end{cases}. \tag{6.7}$$

Let us suppose that the training dataset is a set $\mathcal{T}_{\text{training}} = \{\mathbf{x}_1, \mathbf{x}_2, \dots, \mathbf{x}_{N-M}\}$ and the test dataset is $\mathcal{T}_{\text{test}} = \{\mathbf{x}_{N-M+1}, \mathbf{x}_{N-M+2}, \dots, \mathbf{x}_M\}$. To find the optimum weights $\mathbf{w}$ and threshold θ that correctly classify the M test points in this binary classification problem, we are going to use an error function that calculates if we correctly classify an input or not. An error or cost function that we can use for this can be

$$J = \frac{1}{N - M} \sum_{i=1}^{N-M} (y_* - y)^2 \tag{6.8}$$

or

$$J = \frac{1}{N - M} \sum_{i=1}^{N-M} |y_* - y|, \tag{6.9}$$

where $y_* \in \{-1, +1\}$ is the target function we want to be able to predict and $y \in \{-1, +1\}$ is the predicted function. We will usually prefer the squared error and not the absolute value difference because the squared function is continuous everywhere and we could easily find the minimum of equation (6.8) with respect to the weights $\mathbf{w}$.

Since equation (6.7) and other similar step-functions are discontinuous, we cannot use the derivative to minimize the classification error function with respect to the weights (see the section 6.3.2). We are then obliged to resort to an iterative update weights algorithm to obtain the optimum weights that will define our hyperplane. This algorithm is known as the perceptron training rule or as the Widrow–Hoff algorithm (Rojas 1996).

The Widrow–Hoff algorithm for perceptron weight updating.

---

1. Initialize $\mathbf{w}_1 = 0$.
2. Obtain $x_t \in \mathcal{R}^n$.
3. **For** $t = 1$ to T **do**.
4. Predict $y_t^* \leftarrow \mathbf{w}_t \cdot \mathbf{x}_t$.
5. Find the actual classification y_t.
6. Calculate the error $\Delta y \leftarrow y_t^* - y_t$.
7. Update the weights $\mathbf{w}_{t+1} \leftarrow \mathbf{w}_t - \eta\,(\mathbf{w}_t \cdot \mathbf{x}_t - y_t)\mathbf{x}_t$.
8. **end for**.

As we started with the supposition that the data are linearly separable, we did cheat in this manner, but here the learning algorithm is our only concern. The theorem referring to convergence of this algorithm checks its bounds and may be seen in the paper by Cesa-Bianchi et al (Cesa-Bianchi *et al* 1997).

Assuming convergence we may then use the optimum weights to perform a *forward-pass* with a test input vector $\mathbf{x}_M \in \mathcal{T}_{\text{test}}$. We will assume that linear separability is learned and that it is highly likely that most test inputs will be correctly classified. This algorithm shows how we can use neural networks to learn a linear classifier or regression.

6.3.1 Continuous activation function perceptron

The biggest problem with this perceptron is that it is only a binary classifier that solves linearly separable classifications, even if we make complex connections between multiple perceptrons. The manner in which we can introduce non-linearity and make non-linear classifications as in figure 6.4 is by evaluating $z = \mathbf{w} \cdot \mathbf{x}$ from

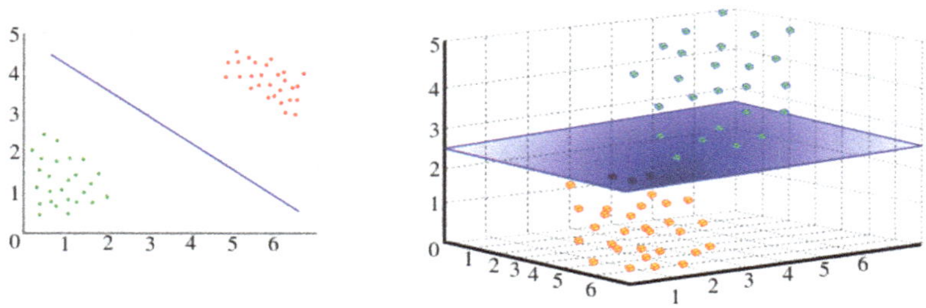

Figure 6.4. A visualization of the sample classifications that an artificial neural network can perform.

equation (6.5) into a non-linear function. We will call this non-linear function an *activation function* for the perceptron. Three commonly used activation functions are the Sigmoid function,

$$\sigma(z) = \frac{1}{1 + e^{-z}}, \tag{6.10}$$

the rectified linear unit (ReLU) function (Arora *et al* 2018),

$$\sigma(z) = \max(0, z) = \begin{cases} z, & \text{if } z \geqslant 0 \\ 0, & \text{if } z < 0 \end{cases}, \tag{6.11}$$

and the hyperbolic tangent function,

$$\sigma(z) = \tanh(z) = \frac{2}{1 + e^{-2z}} - 1. \tag{6.12}$$

Specifically, the ReLU activation function is more widely used in popular neural networks due to that fact that only certain perceptrons will fire and therefore this will be more efficient by performing as many calculations (He *et al* 2020). Using either of the two cases (6.11) or (6.10) as activation functions, we can use gradient descent to find the minimum error in the perceptron output. To further generalize the linear perceptron model we will extend each perceptron to have its own weight b, called bias. Then equation (6.6) can be seen as

$$f(z) = \mathbf{w} \cdot \mathbf{x} + b. \tag{6.13}$$

The perceptron can now take a non-linear form (with an adequate activation function)

$$f(\mathbf{x}, \mathbf{w}, \theta) = \sigma(\mathbf{w} \cdot \mathbf{x}) = \sigma(z). \tag{6.14}$$

We have modified equation (6.6) and we may now minimize the classification error with respect to the weights, using the chain rule:

$$J = \frac{1}{N}(f_* - f)^2 \tag{6.15}$$

$$\frac{\partial J}{\partial \mathbf{w}} = \frac{\partial J}{\partial f}\frac{\partial f}{\partial z}\frac{\partial z}{\partial \mathbf{w}}, \tag{6.16}$$

then

$$\frac{\partial J}{\partial \mathbf{w}} = -\frac{2}{N}(f_* - f)\sigma'(z)\mathbf{x}, \tag{6.17}$$

where $\sigma(z)$ is any continuous function, f_* is the target output and f is the predicted output.

If we use gradient descent we can obtain the change in the weights (also known as the delta rule) $\mathbf{w}$:

$$\mathbf{w}_{t+1} = \mathbf{w}_t + \gamma \nabla J(\mathbf{w}_t), \tag{6.18}$$

where $\gamma \in \mathcal{R}_+$ should be small enough so that $J(\mathbf{w}_t) \geqslant J(\mathbf{w}_{t+1})$ and the weight values converge. Substituting equation (6.17) in the previous equation:

$$\mathbf{w}_{t+1} = \mathbf{w}_t + \eta\left(f_* - f\right)\sigma'(z)\mathbf{x}, \tag{6.19}$$

where $\eta \in \mathcal{R}_+$ is a constant called the *learning rate*, $\nabla J(\mathbf{w}_t) = \frac{\partial J}{\partial \mathbf{w}}$ was used and a negative sign in the second term of the right-hand side of equation (6.19) was added due to our search for a minimization in the J function.

6.3.2 Single perceptron implementation

Next we present an example using `Python` (see listing 6.1) to illustrate how we may find the optimum weights that may correctly classify linearly separable distributed

```python
import pandas as pd
import numpy as np
import matplotlib.pyplot as plt

def read_setosa_versicolor():
    """
    Read Setosa and Versicolor from the Iris data set.
    Returns
    X : (np.array) Training vectors,
        X.shape : [#samples, #features]
    y: (np.array) Target vectors,
        y.shape : [#samples]
    """
    df = pd.read_csv("https://archive.ics.uci.edu/ml/machine-learning-databases/iris/iris.data", header=None)
    y = df.iloc[0:100, 4].values
    y = np.where(y == "Iris-setosa", -1, 1)
    X = df.iloc[0:100, [0, 2]].values
    return X, y

if __name__ == "__main__":
    X, y = read_setosa_versicolor()

    plt.scatter(X[:50, 0], X[:50, 1],  color="red",
        marker="o", label="setosa")
    plt.scatter(X[50:100, 0], X[50:100, 1], color="blue",
        marker="x", label="versicolor")

    plt.xlabel("petal length")
    plt.ylabel("sepal length")
    plt.legend(loc="upper left")
    plt.show()
```

Listing 6.1 Implementation of `read_iris.py`.

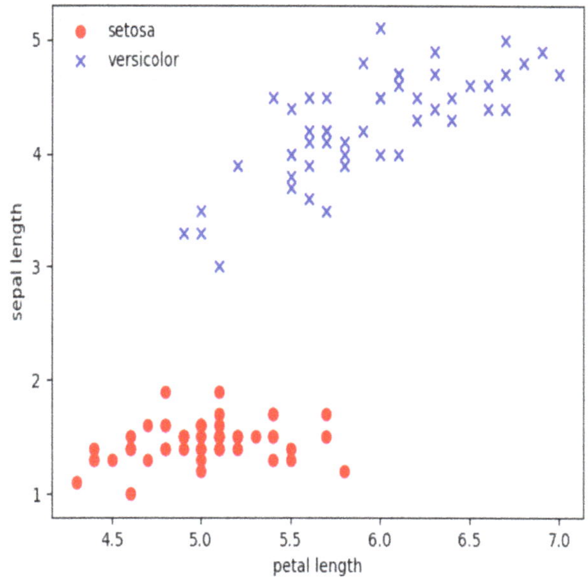

Figure 6.5. Output for listing 6.1: `read_iris.py` main output.

points, as shown in figure 6.5, using the `pandas`, `numpy` and `matplotlib` libraries.

Fisher's Iris dataset consists of 50 samples from each of three species of the iris flower: *Iris setosa*, *Iris virginica* and *Iris versicolor*. Four features were measured from each sample: the lengths and the widths of the sepals and petals, in centimetres.

We will be using a single perceptron (a single layer neural network) to learn how to classify two linearly separable classes on a Cartesian plane. We will only use two features and two flower species for this task. The two linearly separable features are plotted in figure 6.5.

First we create a function that extracts 100 target values `y` stored in the fourth column of the `Pandas` dataframe `df`, then we encode the flower species *Iris setosa* as `-1` and the species *Iris versicolor* as `1`. Next we will extract two feature columns, in this case, the first two features are sepal length and petal length. We extract 100 training samples and name our inputs the `X` `Pandas` dataframe. Finally we use `plt` to visualize our collected training data.

After visualizing our data we construct a *Perceptron class* seen in the following file as `perceptron.py` that uses the Widrow–Hoff algorithm shown in this section and is displayed in the code in listing 6.2 (see figure 6.6 for the output).

Next we will see if our algorithm converges and visualize the boundary decision that our perceptron learned, but first we define the colour mapping in the plot.

We check to see if our algorithm converges in its training with ten epochs and a learning rate of 0.1 (we encourage the reader to play around with these *hyper-parameters* to experiment with different results).

```
import numpy as np

class Perceptron(object):
    def __init__(self, lr = 0.01, iter = 10):
        self.lr = lr
        self.iter = iter

    def fit(self, X, y):
        """Fit training data
        X : Training vectors, X.shape : [#samples, #features]
        y : Target values, y.shape : [#samples]
        """

        self.weight = np.zeros(1 + X.shape[1])
        self.errors = []


        for i in range(self.iter):
            err = 0
            for xi, target in zip(X, y):
                delta_w = self.lr * (target - self.predict(xi))
                self.weight[1:] = self.weight[1:] + delta_w * xi
                self.weight[0] = self.weight[0] + delta_w
                err = err + int(delta_w != 0.0)

            self.errors.append(err)
        return self

    def net_input(self, X):
        return np.dot(X, self.weight[1:]) + self.weight[0]

    def predict(self, X):
        return np.where(self.net_input(X) >= 0.0, 1, -1)
```

Listing 6.2.1 Implementation of `perceptron.py` (auxiliary function).

6.4 Multilayer perceptrons

Let us now consider a case similar to equation (6.5) where more than one perceptron is present. These perceptrons may be in series or in parallel. If they are in parallel they belong to the same layer and if they are in series we will say that the perceptrons belong to different layers. The usual convention is to call the inputs a layer, the corresponding activation functions another layer and the final outputs another layer, such as in figure 6.3. The more layers a multilayer perceptron (MLP) has the *deeper* we will say that it is and those layers are called *hidden layers*.

6.4.1 Backpropagation

These MLPs can introduce more complex non-linearities as they become more complex composite functions as we add more layers. Each perceptron has its own

```python
from matplotlib.colors import ListedColormap
from matplotlib import pyplot as plt
import numpy as np

def plot_decision_regions(X, y, classifier, resolution=0.02):

    """ setup marker generator and color map"""
    markers = ('s', 'x', 'o', '^', 'v')

    colors = ('red', 'blue', 'lightgreen', 'gray', 'cyan')
    cmap = ListedColormap(colors[:len(np.unique(y))])

    """ plot the decision surface """
    x1_min, x1_max = X[:, 0].min() - 1, X[:, 0].max() + 1
    x2_min, x2_max = X[:, 1].min() - 1, X[:, 1].max() + 1

    xx1, xx2 = np.meshgrid(np.arange(x1_min, x1_max, resolution),
    np.arange(x2_min, x2_max, resolution))

    Z = classifier.predict(np.array([xx1.ravel(), xx2.ravel()]).T)
    Z = Z.reshape(xx1.shape)

    plt.contourf(xx1, xx2, Z, alpha=0.4, cmap=cmap)
    plt.xlim(xx1.min(), xx1.max())
    plt.ylim(xx2.min(), xx2.max())

    """ plot class samples """
    for idx, cl in enumerate(np.unique(y)):
        plt.scatter(x=X[y == cl, 0], y=X[y == cl, 1],
        alpha=0.8,
        color=cmap(idx),
        marker=markers[idx],
        label=cl)
```

Listing 6.2.2 Implementation of `color_map_perceptron.py` (auxiliary function).

activation function, weights, bias, inputs and outputs. As this is the case, there may be a number (l) of layers and a number of perceptrons per layer (n) that we must generalize so that the algorithm finds the optimum weights.

Backpropagation is the name of the algorithm that minimizes the error function in the weight space using gradient descent throughout the neural network. It is a two-step iterative process:

1. Initially, we can choose a random $\mathbf{w}$ weight matrix that represents the weight vector input in every layer. Usually we choose `small` values such as $\mathbf{w} \in [0, 1]$
2. We then forward-propagate the initial inputs $\mathbf{x}$ through the whole network and we calculate the average error output in the final layer.
3. Next, we backward-propagate this error and update the weight matrix $\mathbf{w}$ with equation (6.19).

```
from matplotlib import pyplot as plt

from read_iris import read_setosa_versicolor
from perceptron import Perceptron
from color_map_perceptron import plot_decision_regions

X, y = read_setosa_versicolor()
tron = Perceptron(0.1, 10)
tron.fit(X, y)

plt.plot(range(1,len(tron.errors)+1),tron.errors,marker="o")
plt.xlabel("Epochs")
plt.ylabel("Number␣of␣misclassifications")
plt.show()

plot_decision_regions(X, y, classifier=tron)
plt.xlabel("sepal␣length␣[cm]")
plt.ylabel("petal␣length␣[cm]")
plt.legend(loc="upper␣left")
plt.show()
```

Listing 6.2.3 Implementation of `plot_epochs_boundary.py`.

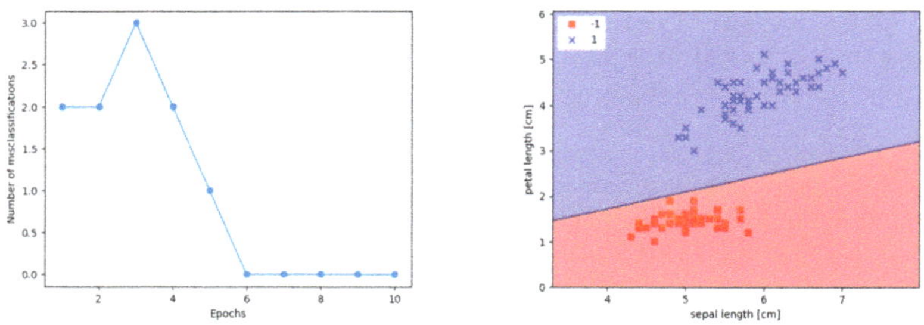

Original Image | Processed New Image

Figure 6.6. Output for listing 6.2: plotting the epoch boundary.

4. Then we repeat steps 2 and 3 until we obtain a convergence for the weight matrix **w**. Steps 2 and 3 are called *forward propagation* and *backward propagation*, respectively, and are seen as one iteration called an *epoch*.

There exist more options than starting the weight matrix with initially random values $\mathbf{w} \in [0, 1]$ depending on the optimization problem (see more on transfer learning) but this is a universal starting point.

We may also use more or fewer data inputs per training epoch. Each of these inputs may be seen as training samples. We call these inputs *batches* of data and there is a trade-off in computer memory utilization with respect to minimization error when using different batch sizes.

In regular *gradient descent* (GD), the error is summed over all training examples before updating weights, whereas in *stochastic gradient descent* (SGD) weights are updated upon examining each training example.

Summing over multiple examples in standard gradient descent requires more computation per weight update step.

On the other hand, because it uses the true gradient, standard GD is often used with a larger step size per weight update than SGD. In cases where there are multiple local minima with respect to the error J, SGD can sometimes avoid falling into these local minima.

Similar to a Fourier series, the more terms included in an expansion, or in this case, the deeper our neural network, the better the approximation to our non-explicit function f (assuming global minima convergence in the equation (6.20)).

In the `Python` example of listing 6.3 we use an MLP to learn the XOR. We can change the activation function and its corresponding derivative as long as the

```
import numpy as np

""" Activation function and its derivative """
def tanh(x):
    return (1.0 - np.exp(-2*x))/(1.0 + np.exp(-2*x))

def tanh_derivative(x):
    return (1 + x)*(1 - x)

class NeuralNetwork:

    def __init__(self, net_arch):
        """
        net_arch: consists of a list of integers,
          indicating
              the number of neurons in each layer
        """
        np.random.seed(0)

        """Initialize the weights and biases """
        self.activity = tanh
        self.activity_derivative = tanh_derivative
        self.layers = len(net_arch)
        self.steps_per_epoch = 1
        self.arch = net_arch
        self.weights = []

        """ Random initialization with range of weight
          values (-1,1)"""
        for layer in range(self.layers - 1):
            w = 2*np.random.rand(net_arch[layer] + 1,
              net_arch[layer+1]) - 1
            self.weights.append(w)
```

Listing 6.3 Auxiliary functions of `MultiLayeredPerceptron.py`.

```python

    def _forward_prop(self, x):
        y = x

        for i in range(len(self.weights)-1):
            activation = np.dot(y[i], self.weights[i])
            activity = self.activity(activation)

            """ add the bias for the next layer """
            activity = np.concatenate((np.ones(1), np.array(activity)))
            y.append(activity)

        """ last layer """
        activation = np.dot(y[-1], self.weights[-1])

        activity = self.activity(activation)
        y.append(activity)

        return y

    def _back_prop(self, y, target, learning_rate):
        error = target - y[-1]
        delta_vec = [error * self.activity_derivative(y[-1])]

        """ we begin at last layer"""
        for i in range(self.layers-2, 0, -1):
            error = delta_vec[-1].dot(self.weights[i][1:].T)
            error = error*self.activity_derivative(y[i][1:])
            delta_vec.append(error)

        """ now we set the values from back to front"""
        delta_vec.reverse()

        """ we update the weights using backpropagation """
        for i in range(len(self.weights)):
            layer = y[i].reshape(1, self.arch[i]+1)
            delta = delta_vec[i].reshape(1, self.arch[i+1])
            self.weights[i] += learning_rate*layer.T.dot(delta)

    def fit(self, data, labels, learning_rate=0.1, epochs=100):
        """
        data:    set of all possible pairs of 1 or 0
        labels:  result of 'xor' on each of data pairs
        """

```

Listing 6.3 (Continued.)

```
        """ Add bias units to the input layer -
        add a "1" to the input data """
        ones = np.ones((1, data.shape[0]))
        Z = np.concatenate((ones.T, data), axis=1)

        for k in range(epochs):
            sample = np.random.randint(data.shape[0])

            """ we set up our feed-forward propagation """
            x = [Z[sample]]
            y = self._forward_prop(x)

            """ we do back-propagation of the error to
              adjust the weight"""
            target = labels[sample]
            self._back_prop(y, target, learning_rate)

    def predict_single_data(self, x):
        """ to check the single prediction result
        x:      single input data   """


        val = np.concatenate((np.ones(1).T, np.array(x)))
        for i in range(0, len(self.weights)):
            val = self.activity(np.dot(val,
                           self.weights[i]))
            val = np.concatenate((np.ones(1).T,
                           np.array(val)))
        return val[1]

    def predict(self, X):
        Y = np.array([]).reshape(0, self.arch[-1])
        for x in X:
            y = np.array([[self.predict_single_data(x)]])
            Y = np.vstack((Y,y))
        return Y
```

Listing 6.3 (Continued.)

variable names are changed in the neural network class. Then we define the neural network class which works with activation functions.

Using our neural network class (which uses the `tanh` and `tanh_derivative` functions), we initialize our input data with an architecture that has two input neurons, two hidden neurons and one output neuron.

We then train and predict to obtain an XOR approximation (see listing 6.4 for the code, and figures 6.7 and 6.8 for the output).

We can use different architectures in order to visualize different boundary decisions.

```python
from MultiLayeredPerceptron import NeuralNetwork
import sys, os
sys.path.append(os.path.abspath(".."))
from listing_6_01.color_map_perceptron import plot_decision_regions
from matplotlib import pyplot as plt
import numpy as np

np.random.seed(0)
nn = NeuralNetwork([2,2,1])
 # may also try NeuralNetwork([2,3,4,1])

X = np.array([[0, 0], [0, 1], [1, 0], [1, 1]])
y = np.array([0, 1,  1, 0])

nn.fit(X, y, epochs=10000)

print("Predictions")
print()
for s in X:
    print("XOR point [x y]: ",s," Classified as: ", nn.predict_single_data(s))

plot_decision_regions(X, y, nn)
plt.xlabel("x-axis")
plt.ylabel("y-axis")
plt.legend(loc="upper left")
plt.tight_layout()
plt.show()
```

Listing 6.4 Implementation of `show_predictions.py`.

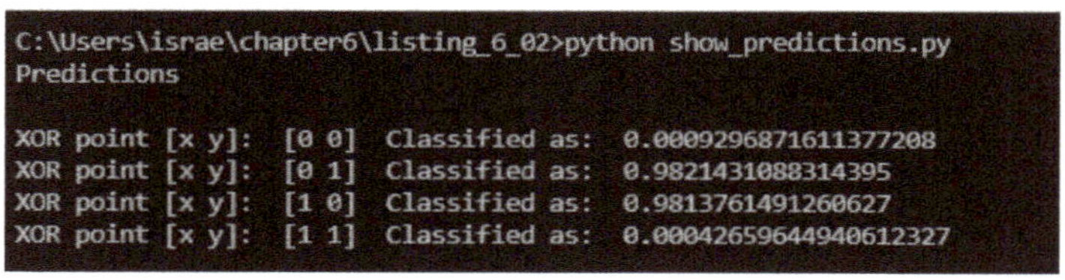

Figure 6.7. Output for listing 6.4: `Show_predictions.py` terminal output.

6.4.2 Maximum likelihood—binary cross-entropy

To better generalize the training data and to make a neural network successfully perform on unseen data (called *validation data*) we can use a popular error function called the *cross-entropy function* that will let us know how well we classified our data on the forward propagation. Statistically, maximum likelihood will let us know what class a sample input most probably belongs to with certain parameters. We may write the maximum likelihood estimate (MLE) as

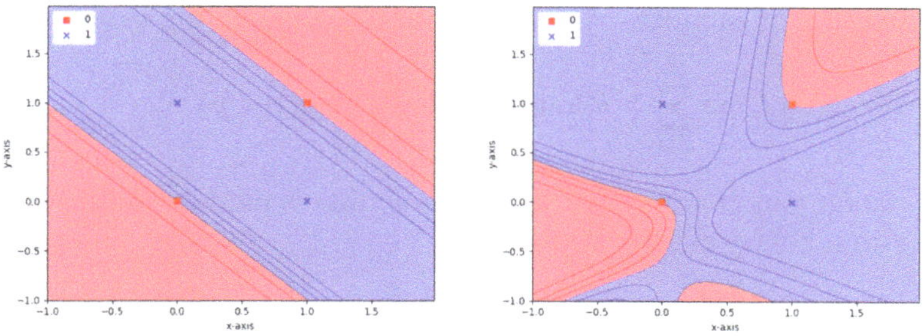

3-layer Network (2, 2, 1) perceptrons   4-layer Network (2, 3, 4,1) perceptrons

Figure 6.8. Output for listing 6.4: learn the XOR boundary for a neural network.

$$\hat{\mathbf{w}} = \arg\max(L(\mathbf{x}|\mathbf{w})), \tag{6.20}$$

where the likelihood $L(\mathbf{x}|\mathbf{w})$ is given by the product of all possible probabilities which is called the *joint probability* (the probability of all probabilities occurring)

$$L = \Pi_{i=1}^{N} P(\mathbf{x}|\mathbf{w}), \tag{6.21}$$

where $P(\mathbf{x}|\mathbf{w})$ is the probability of $\mathbf{x}$ given $\mathbf{w}$. Likelihood measures the goodness of fit of a statistical model to a sample of data for given values of the unknown parameters. Due to numerical underflow while computing the joint probability, we can use the log of equation (6.21) to compute sums of small quantities instead of products. By using the log function of the likelihood the location of the maximum of L will not change as the logarithm function is strictly increasing and so if $x > y$ then $\log(x) > \log(y)$ always, and so maximum values are maintained for our weights $\mathbf{w}$:

$$L_* = \log(L) = \sum_{i=1}^{N} \log(P(\mathbf{x}|\mathbf{w})). \tag{6.22}$$

For computational standards we also usually minimize the error functions and so instead of calculating the maximum of the log-likelihood we can calculate the minimum of the negative log-likelihood as

$$\hat{\mathbf{w}} = \arg\min(-L_*). \tag{6.23}$$

To see how this can happen in a binary classification let us return to a simple binary classification example.

Say we have only two events A and B where the probability of obtaining event A randomly is p and the probability of obtaining B randomly is $1 - p$ and we have n trials, then we can say that these events follow a *Bernoulli probability distribution* represented as

$$B = \begin{Bmatrix} p & \text{if A} \\ 1 - p & \text{if B} \end{Bmatrix} = p^n(1 - p)^{1-n} \tag{6.24}$$

for any n number of samples. Next we obtain the negative log-likelihood as

$$-L_* = \sum_{n=1}^{N} - n \log(p) + (1 - n) \log(1 - p). \tag{6.25}$$

This is the famous *cross-entropy loss function* that is used to find the highest class probability in a binary classification.

6.4.3 Maximum likelihood—multiple category cross-entropy

To generalize the loss function to solve a *multi-class single label* problem. We will use as an auxiliary a different activation function in the last layer of a multilayer perceptron (LeCun 1988).

If we used the so-called *Softmax function* or *normalized exponential function* as

$$S(\mathbf{z}) = \frac{e^{z_j}}{\sum_{j}^{K} e^{z_j}}, \tag{6.26}$$

where each z_j for $j = 1, 2, 3, \ldots, K$ are the outputs of each perceptron in the final layer. We can normalize the final output to a probability distribution over the predicted output classes. We will call this activation function in conjunction with a generalized non-binary cross-entropy loss function the *categorical cross-entropy function*:

$$L_{M_{\text{class}}} = \sum_{m=1}^{M} - n_m \log(p_m). \tag{6.27}$$

We can use this cross-entropy function when there are more than two classes $M > 2$ in a single label classification because the probability output is normalized and each class is mutually exclusive.

If we were to generalize the problem even more, as seen in the ImageNet competitions (Krizhevsky *et al* 2017), then we would have to solve a multi-label multi-class classification. This can be achieved by resorting to a non-normalized final output vector. Such an output can be achieved by using a sigmoid, ReLu or hyperbolic tangent activation function in the final layer, and computing the loss by using the binary cross-entropy function. This will give us the probability of a certain label taking into account all of the classes.

6.5 Convolutional neural networks

6.5.1 Introduction

In the last section we saw MLPs, which we can refer to as neural networks (NNs), and we saw how they use a dataset of training data as an input, how they are trained for generalization with backpropagation, which types of problems they can solve and how we can make inferences (or *guesses*) for unseen data. We will go one step further now and see how convolutional neural networks work (Zeiler and Fergus 2014) , and why we use them specifically for image classification, although they can also serve for text classification or audio recognition.

6.5.2 Convolution and cross-correlation

We will begin by introducing the 1D convolutional operation which is commonly seen in signals and with continuous functions:

$$f(x)*g(x) = \int_{-\infty}^{\infty} f(\tau)g(x - \tau)d\tau, \tag{6.28}$$

where $f*g$ is read as f convolution g. In the discrete function case (which we can handle more conveniently due to numerical computation) this can be seen in 1D and 2D, respectively, as

$$f[x]*g[x] = \sum_{k=-\infty}^{\infty} f[k]g[x - k] \tag{6.29}$$

$$f[x, y]*g[x, y] = \sum_{k_1=-\infty}^{\infty} \sum_{k_2=-\infty}^{\infty} f[k_1, k_2]g[x - k_1, y - k_2], \tag{6.30}$$

where $f[x, y]$, $g[x, y]$ are discrete 2D functions, k corresponds to the kernel size in 1D and k_1, k_2 correspond to the kernel size matrix in 2D. We see the 2D discrete convolution as the result of the repeated sum of different products operating over a discrete function f with a discrete function g that is mirrored in its rows and columns. We take into account that if the kernel g has a number of odd elements, the centre or origin $x = 0$, $y = 0$ of the function is located in the centre column of the centre row.

The convolution is an operation that basically sums the product of two functions over a certain dimension. In 2D this can be visualized as the sum of the product of two matrices while one matrix *slides* over the other as seen in figure 4.1.

Similar to the convolution in 2D, cross-correlation is defined as

$$f[x, y]*g[x, y] = \sum_{k_1=-\infty}^{\infty} \sum_{k_2=-\infty}^{\infty} f[k_1, k_2]g[x + k_1, y + k_2]. \tag{6.31}$$

If we were to compare cross-correlation and convolution, the only difference between the two operations is a *mirror flip* that may be seen in the order of the elements of the matrix G which corresponds to the function g. Supposing G is of dimensions $M \times N$,

$$\sum_{i=1}^{M} \sum_{j=1}^{N} G_{ij} \to \sum_{i=M}^{1} \sum_{j=N}^{1} G_{ij} \tag{6.32}$$

occurs before the *sliding window* operation in a convolution.

In convolutional neural networks (CNNs), in each convolutional layer the weights are stored in a g discrete function that operates over an input f discrete function. These weight matrices are popularly called *filters*. In numerical computations we may

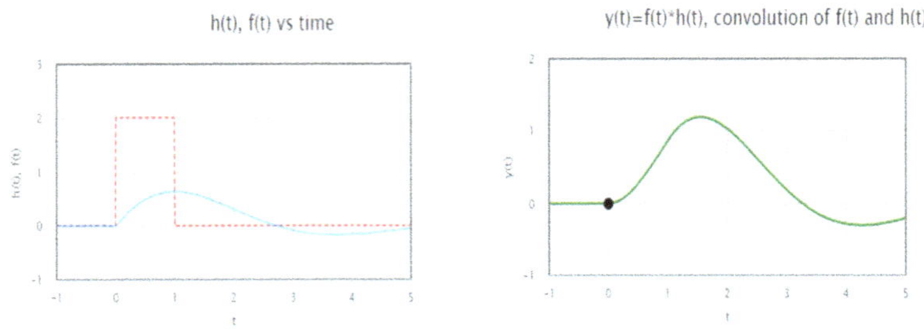

Square wave over a damped Sine function. Convoluting it with a 2D square wave.

Figure 6.9. An example after convolution with a 2D square wave and sine wave.

implement the use of cross-correlations instead of convolutions as many machine learning libraries do.

Due to its similarity with convolution, its simplicity to compute in backpropagation, and the fact that regardless of whether convolution or cross-correlation is performed, the weights will be learned correspondingly in the final process.

We can see that by convoluting a square wave over a damped sine function we may obtain a function that contains the basic information of the damped sine function (indeed this can act as an averaging or blur filter if we extrapolate this result to a 2D image). We may see that the convolution of an image with 2D square impulse results in a sharpening effect. We may assume that any local spatial feature from a greyscale image may be extracted with a cross-correlation or convolution with translational invariance (O'Shea and Nash 2015). See figure 6.9.

6.5.3 Why CNNs instead of MLPs?

Technically, everything that an MLP can do, a CNN can do as well. But let us talk in terms of convenience, if we choose to forward-propagate a greyscale image ($M \times N$) through an MLP we must first flatten the image into an input vector and must include as many layers as necessary to be able to capture local spatial features to finally generalize to unseen images with cross-entropy loss. There may be some problems here with locality and translation of key features which may be lost and the amount of weights and operations that are needed in order to have a well-trained MLP are usually larger than a CNN which makes overfitting more probable.

Let us compare the efficiency of both a CNN and an MLP over an $M \times N \times C$ tensor or image. For an n_{output}-class classification:

1. For a single layer MLP, we have that the number of trainable parameters will be $n_{\text{parameters}} = (M \times N \times C_{\text{in}}) \times n_{\text{output}}$.
2. For a single layer CNN, we have that the number of trainable parameters will be $n_{\text{parameters}} = (k_x \times k_y \times C_{\text{in}} \times C_{\text{out}})$, where c_{out} is the number of filters and k_x, k_y represent the kernel width and height, respectively.

Since CNNs depend on the kernel size, where usually the convolutional kernel size is smaller than the input image dimensions that it is convoluting, the amount of trainable parameters is reduced while extracting spatial features at the same time. Here c_{out} will create a *deeper* network and we will be able to extract different features with each filter.

When numerically computing the *convolution* that is actually a cross-correlation we have four factors, or so-called *hyperparameters*, that we may adjust:

1. The size of the stride S when *sliding* the kernel over the image.
2. The number of filters F that we want to use to cross-correlate.
3. The size of the kernel $K = \dim k_x \times k_y$.
4. Padding P around the image to control the output size. Two types of padding P are shown in figure 6.10.

Hyperparameters are called such because they are adjusted heuristically and are not inherent to the input data. They are different from other parameters because hyperparameters cannot be learned by the CNN model. A list of hyperparameters for a CNN include, but are not limited to, the number of layers, activation functions, number of filters, size of filters, sliding window size and learning rate in backpropagation.

A square kernel $K = \dim k_x \times k_y \in \mathcal{Z}^+$ is usually used because it is numerically and computationally practical and serves as a clean canvas to learn any symmetrical or non-symmetrical features (such as edges, curves, sharpness, etc), they are also minimally biased unlike rectangular or other shaped kernels.

Assuming a square kernel, a convolutional layer then accepts an input tensor of dimensions

$$M \times N \times C_{\text{in}} \tag{6.33}$$

Figure 6.10. An example of padding in a discrete 2D convolution operation. Credit: Highsmith, Carol M., 1946 - United States Library of Congress's Prints and Photographs Division under the digital ID highsm.09920.

and outputs a tensor of dimensions

$$W \times H \times C_{\text{out}}, \tag{6.34}$$

where $W = (M - F + 2P)/S + 1$, $H = (N - F + 2P)/S + 1$ and $C_{\text{out}} = K$.

Similar to MLPs we use an element-wise activation function after the convolutional layer to obtain non-linearity. Another part of a CNN is a called *pooling layer*, which downsamples the dimensionality of feature maps which further reduces overfitting and computational cost. It does this by replacing the output at a certain location with a statistic of the nearby outputs. An example of pre-pooled feature maps may be seen in figure 6.11 where we may visualize the abstract layers that a CNN learns to predict different classes.

The pooling layers used are usually one of two main types, max pooling or average pooling, which can be seen in figure 6.12, although many others exist, such as the L2 rectangle neighbourhood or a weighted average based on central pixel

Image resource[a].

[a]Courtesy of website: convnetplayground.fastforwardlabs.com/#/

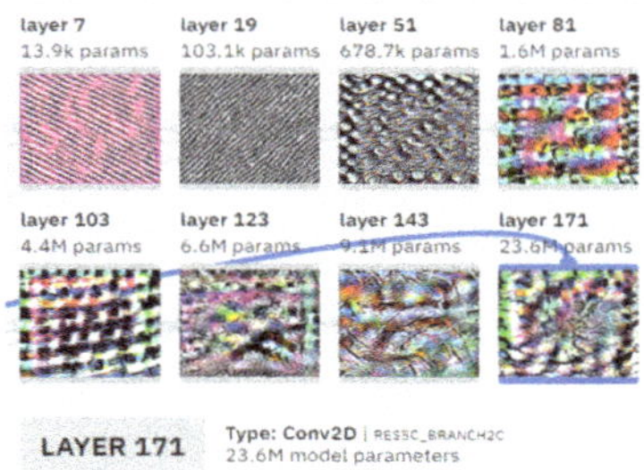

Image process.

Figure 6.11. Example of an image of a boat from CIFAR-10 and several activation maps of ResNet-50.

Figure 6.12. Examples of pooling in a discrete 2D operation. Credit: Highsmith, Carol M., 1946 - United States Library of Congress's Prints and Photographs Division under the digital ID highsm.09920.

distance. In general, pooling layers may be substituted by diluted convolutions but this is beyond the scope of this book.

CNNs typically rely on the fact that we will substitute MLP neuron layers for convolutional layers and pooling layers, except for the final layer, in order to be able to make a final classification decision. Also a common practice is to include *dropout* in the CNN layers, such that overfitting becomes seemingly less plausible. Dropout is a technique that randomly deactivates neurons in the backpropagation process.

As an example, we use a popular dataset, the handwriting MNIST dataset, where we will use a CNN to correctly classify the images of numbers (see figure 6.13). For time optimization purposes we will use the Keras library which works on top of Tensorflow, which was developed by Google. CNN models where the layers are sequential are popular because of their functionality, but we remind the reader that these are not the only CNN architectures that may exist.

We build our CNN model with two convolutional layers, where we have 32 and 64 weight matrices of size 3 × 3 with ReLU activation for each layer. This is followed by a max pooling layer using a small dropout. In the end we use a *fully connected layer* which is basically a continuous perceptron that has 128 inputs and ten outputs. From the Keras documentation: '*Dense implements the operation: output = activation(dot(input, kernel) + bias) where activation is the element-wise activation function passed as the activation argument, kernel is a weights matrix created by the layer, and bias is a bias vector created by the layer…*' https://faroit.com/keras-docs/2.0.6/layers/core/.

We train this model with a batch size of 128 (meaning 128 × 28 × 28 = 100 Kb are stored in the memory for each epoch) and SGD is used with the Keras standard `learning_rate = 0.01`. More optimizers exist such as Adagrad, Adam, RMSprop and others, which can be found in the Keras documentation. These other optimizers use different methods to minimize the loss function with respect to

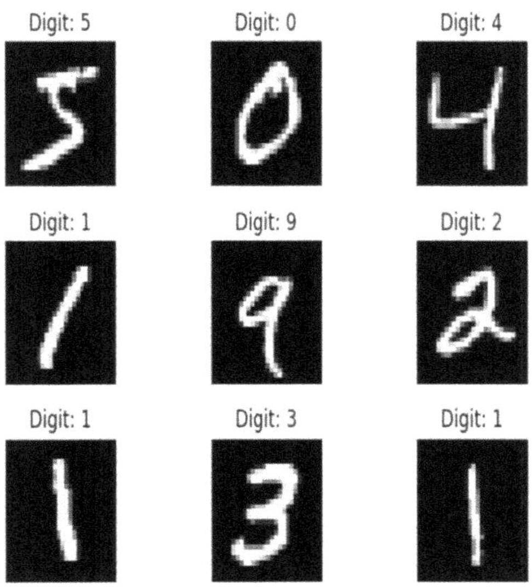

Figure 6.13. `train_cnn.py` MNIST samples.

the weights and often have a changing learning rate parameter called *momentum*. Finally we plot our CNN metrics in the training and testing epochs (see listing 6.5 for the `Python` code and figures 6.14 and 6.15 for the output).

Effectively, the tensor size output, total weights (parameters) and memory requirements of each layer when an image of size 28 × 28 × 1 is passed through this CNN architecture would be as follows:

1. Input layer: [28 × 28 × 1]; memory: 28 × 28 = 784 K; weights: 0
2. Convolutional layer (32 filters of 3 × 3): [28 × 28 × 32]; memory: 28 × 28 × 32 = 25.1 K; weights: (3 × 3 × 1 + 1) × 32 = 320.

```python

import numpy as np
import matplotlib
#matplotlib.use("agg") # if usinf Linux
#matplotlib.use("Qt5Agg") # if using Conda env w/ pip install PyQt5
#matplotlib.use("TkAgg")# if using Ubuntu env
import matplotlib.pyplot as plt

import os

from keras.datasets import mnist
from keras.models import Sequential, load_model
from keras.layers.core import Dense, Dropout, Activation, Flatten
from keras.layers import Conv2D, MaxPooling2D
from keras.utils import np_utils
from keras import backend as k

import tensorflow as tf

""" split the data set into training and testing data """
(X_train, y_train), (X_test, y_test) = mnist.load_data()

""" we see some figures of the data set.
This MNIST data set comprises 60000 images of 10 classes
of the numbers 0 to 9 in a 28x28x1 format """

for i in range(9):
  plt.subplot(3,3,i+1)
  plt.tight_layout()
  plt.imshow(X_train[i], cmap="gray", interpolation="none")
  plt.title("Digit: {}".format(y_train[i]))
  plt.xticks([])
  plt.yticks([])
plt.show()
""" assumes our data format
 "channels_last" assumes (conv_dim1, conv_dim2, conv_dim3, channels) """
```

Listing 6.5 Implementation of `train_cnn.py`.

```
"""
 "channels_first" assumes (channels, conv_dim1, conv_dim2, conv_dim3)
"""

img_rows, img_cols = 28, 28

if k.image_data_format() == "channels_first":
    X_train = X_train.reshape(X_train.shape[0], 1, img_rows, img_cols)
    X_test = X_test.reshape(X_test.shape[0], 1, img_rows, img_cols)
    input_shape = (1, img_rows, img_cols)
else:
    X_train = X_train.reshape(X_train.shape[0], img_rows, img_cols, 1)
    X_test = X_test.reshape(X_test.shape[0], img_rows, img_cols, 1)
    input_shape = (img_rows, img_cols, 1)

"""reshaping"""
X_train = X_train.astype("float32")
X_test = X_test.astype("float32")
X_train = X_train/255
X_test = X_test/255

print("X_train shape:", X_train.shape) #X_train shape: (60000, 28, 28, 1)
num_category = 10

"""convert class vectors to binary class matrices"""
y_train = tf.keras.utils.to_categorical(y_train, num_category)
y_test = tf.keras.utils.to_categorical(y_test, num_category)

model = Sequential()

""" convolutional layer with 32 weight matrices and"""
"""rectified linear unit activation (default: stride = 1, and no padding)"""
model.add(Conv2D(32, kernel_size=(3, 3),
                 activation="relu",
                 input_shape=input_shape))

""" convolutional layer with 64 weight matrices and
rectified linear unit activation """
model.add(Conv2D(64, (3, 3), activation="relu"))

```

Listing 6.5 (Continued.)

```
""" do max pooling (default: strides = pool_size)"""
model.add(MaxPooling2D(pool_size=(2, 2)))

""" randomly turn neurons off to improve convergence
and reduce overfitting """
model.add(Dropout(0.25))

""" flatten dimensions for a
multiclassification output """
model.add(Flatten())

""" fully connected network """
model.add(Dense(128, activation="relu"))
model.add(Dropout(0.5))

"""use a softmax function """
model.add(Dense(num_category, activation="softmax"))

model.compile(loss=tf.keras.losses.categorical_crossentropy
    ,
              optimizer=tf.keras.optimizers.SGD(),
              metrics=["accuracy"])
num_epoch = 10
batch_size = 256

"""model training"""
model_log = model.fit(X_train, y_train,
          batch_size=batch_size,
          epochs=num_epoch, verbose=1,
          validation_data=(X_test, y_test))
score = model.evaluate(X_test, y_test, verbose=0)
print("Test␣loss:", score[0])
print("Test␣accuracy:", score[1])
fig = plt.figure()
plt.subplot(2,1,1)
plt.plot(model_log.history["accuracy"])
plt.plot(model_log.history["val_accuracy"])
plt.title("model␣accuracy")
plt.ylabel("accuracy")
plt.xlabel("epoch")
plt.legend(["train", "test"], loc="lower␣right")
plt.subplot(2,1,2)
plt.plot(model_log.history["loss"])
plt.plot(model_log.history["val_loss"])
plt.title("model␣loss")
plt.ylabel("loss")
plt.xlabel("epoch")
plt.legend(["train", "test"], loc="upper␣right")
plt.tight_layout()
plt.show()
```

Listing 6.5 (Continued.)

```
model_digit_json = model.to_json()
with open("model_digit.json", "w") as json_file:
    json_file.write(model_digit_json)
model.save_weights("model_digit.h5")
print("Saved␣model␣to␣disk")
```

Listing 6.5 (Continued.)

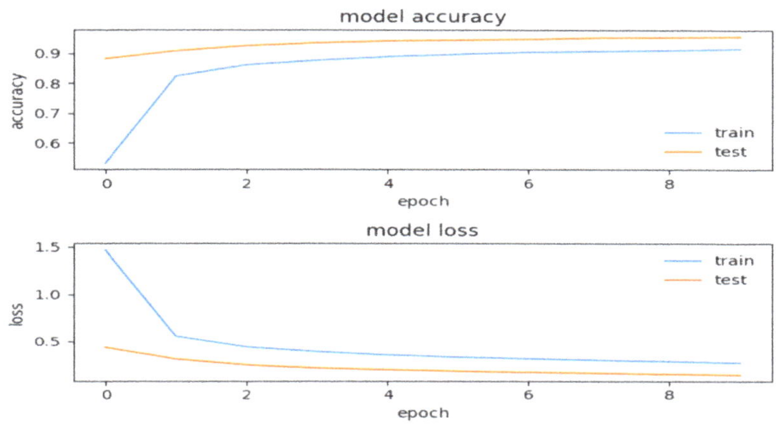

Figure 6.14. `train_cnn.py` accuracy and loss graphs for training and validation using ten epochs over MNIST.

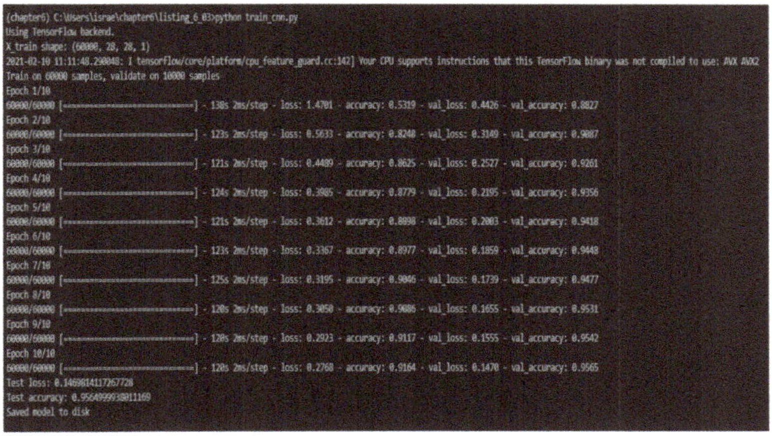

Figure 6.15. `train_cnn.py` ten epoch MNIST training terminal output.

3. Convolutional layer (64 filters of 3 × 3): [28 × 28 × 64]; memory: 28 × 28 × 64 = 50.1 K; weights: (3 × 3 × 32 + 1) × 64 = 18 496.
4. Max pooling layer (2 × 2, stride = 1): [14 × 14 × 6]; memory: 14 × 14 × 64 = 12.5 K; weights: 0.
5. Dropout layer: [14 × 14 × 64]; memory: 14 × 14 × 64 = 12.5 K; weights: 0.

6. Flatten layer: [1 × 12 544]; memory: 1 × 12 544 = 12.5 K; weights: 64 × 12 544 = 802 816.
7. Dense or fully connected layer: [1 × 128]; memory: 128 K; weights: 128 × 12 544 = 1605 632.
8. Dropout layer: [1 × 128]; memory: 128 K; weights: 0.
9. Dense or fully connected layer: [1 × 10]; memory: 10 K; weights: 128 × 10 = 1280.

Here we used `num_classes = 10` for the final layer. The total memory used per forward-pass is around 113.7 K × 4 bytes ≈455 Kb and for a full epoch the memory requirement would be ≈910 Kb; both are considering a batch size of one image. These memory requirements are for calculating a large enough batch size to optimize the training time. Most modern laptops can easily surpass these memory requirements which is why most readers may use a batch size of 256 ≈29 Mb while training this CNN. Using a GPU, we may allocate this memory to realize operations in parallel and have an even faster training time. The total number of parameters is 2428 544 ≈ $2.4M$. The parameters for convolutional layers are calculated using `total_params = (filter_height * filter_width * input_image_channels + 1) * number_of_filters` and for dense layers (without bias) `total_params = input_image_channels * output_image_channels`. The number of parameters is the usual indicator of how fast or slow an inference (forward-pass) can be made using a certain architecture. For example, a very fast inference and compact network architecture, MobileNetV2, has around $3.4M$ parameters, while a deeper and (usually) more accurate network, ResNet-101, has around $44.6M$ parameters.

6.6 Metrics

In a classification task we usually obtain a classification matrix called a *confusion matrix* that gives us information about how the system performed during the training or validation tasks. In a binary classification task a confusion matrix can be seen as a matrix of true positives (TP), true negatives (TN), false positives (FP) and false negatives (FN).

An example of a binary confusion matrix can be seen in figure 6.16. The most popular metrics that can be obtained from this binary confusion matrix can be defined given TP, TN, FP and FN. These metrics include but are not limited to:

- *Accuracy*: $\frac{\mathrm{TP}+\mathrm{TN}}{\mathrm{TP}+\mathrm{TN}+\mathrm{FP}+\mathrm{FN}}$
  is the most popular metric and denotes the weighing of the correctly classified objects over the total number of classified objects.
- *Precision*: $\frac{\mathrm{TP}}{\mathrm{TP}+\mathrm{FP}}$,
  when predicting the positive class this metric tells us how often it is correctly classified.
- *Recall*: $\frac{\mathrm{TP}}{\mathrm{TP}+\mathrm{FN}}$, when the actual class is the positive class, this metric tells us how often it is correctly classified.

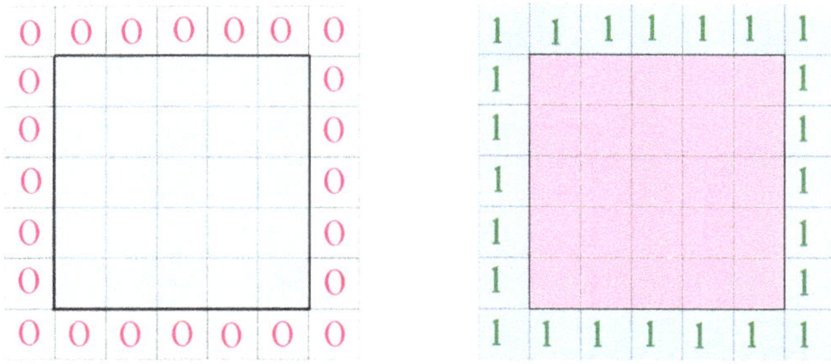

Figure 6.16. Example of a binary confusion matrix.

- *F_1-score*: $2\,\frac{\text{precision} \times \text{recall}}{\text{precision} + \text{recall}}$
  gives the harmonic mean between the precision and the recall.
- *Specificity*: $\frac{\text{TN}}{\text{TN} + \text{FP}}$,
  also called the true negative rate (TNR), gives information about how many of the whole negative class have been classified as negative.
- *False discovery rate* (FDR): $\frac{\text{FP}}{\text{FP} + \text{TP}}$
  gives us information on how many predictions out of all the positive predictions were incorrect.
- *Negative predictive value* (NPV): $\frac{\text{TN}}{\text{TN} + \text{FN}}$
  measures how many predictions out of all the negative predictions were correct.
- *False positive rate* (FPR): $\frac{\text{FP}}{\text{FP} + \text{TN}}$
  measures how wrongly we are classifying the positive class in terms of the other class that was correctly classified.
- *False negative rate* (FNR): $\frac{\text{FN}}{\text{FN} + \text{TN}}$
  measures how wrongly we are classifying the negative class in terms of the other class that was correctly classified.
- *Receiving operator characteristic* (ROC) *curve*:
  This is the plot of the TPR versus the FPR. The random guess curve classification is equivalent to a $y = x$ curve in the first quadrant.
- *ROC area under curve* (AUC) *score*:
  This is then a measure of the performance of classification, where it must be $>\frac{1}{2}$ to be better than random guessing.
- *Precision–recall curve*:
  Having a recall of 1 is not necessarily good as this only indicates that there was not much data classified as the negative class. Usually this means that the precision will also be low, as FPs will be higher. To illustrate this trade-off we can use this curve, as seen in figure 6.17.
- *Log loss*:
  $L = -(y_{\text{true}} \log(y_{\text{pred}}) + (1 - y_{\text{true}})\log(1 - y_{\text{pred}}))$

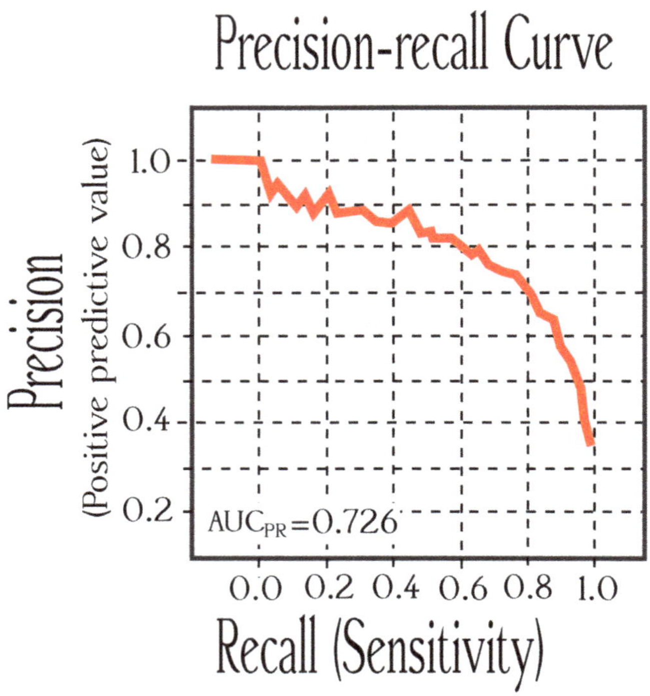

Figure 6.17. Example of a precision–recall curve. ROC curves are similar in nature but with a vertical mirror flip and differently named axes.

is a commonly used cost function that can also be used as a performance metric. The lower the value for an unseen pair (y_{true}, y_{pred}), the better.

A few less commonly implemented classification metrics include:

- *F_β-score*:
  $F_\beta = (1 + \beta)\, \frac{\text{precision} \times \text{recall}}{\beta^2 \times \text{precision} + \text{recall}}$,
  where β acts as a weight, where the higher the importance of recall over precision, the higher the score of $0 < \beta < 1$ you should choose. If $\beta = 1$ we obtain the F_1-score.
- *F_2-score*: This is the special case of F_β using $\beta = 2$. It places double the emphasis on recall.
- *Cohen kappa*:
  The $\kappa = \frac{p_0 - p_e}{1 - p_e}$,
  where p_0 is the observed agreement and p_e is the expected agreement. Observed agreement is how the classifier agrees with the ground truth and the expected agreement is how a random classifier agrees with the ground truth. These two agreements are basically accuracy and random accuracy.

- *Matthew's correlation coefficient* (MCC):
  $$\frac{TP \times TN - FP \times FN}{(TP + FP)(TP + FN)(TN + FP)(TN + FN)}$$
  is the correlation between y_{true} and y_{pred}.
- *Kolmogorov–Smirnov* (KS) *plot and statistic*:
  The KS statistic is the maximum difference between the TPR and the FPR. The KS plot is obtained by taking different probability threshold values and plotting the TP and TN versus threshold probability.
- *Cumulative gains chart*: shows the percentage of the overall number of cases in a given category 'gained' by targeting a percentage of the total number of cases.
- *Gini coefficient*:
  $$G = \frac{AUROC - (1/2)}{1/2} = 2AUROC - 1$$
  is the ratio between the area of the ROC curve and the random model line, and the top left area of the triangle above the random model line. The higher the Gini coefficient, the closer the model is to an ideal classifier.
- *Brier score* :$(y_{pred} - y_{true})^2$ is very similar to the mean squared error and its purpose is to measure differences in the probability space.

In all of these cases we usually define the cut-off probability or threshold probability as $p = 1/2$, which defines the probability mapping to define the classification of one class over the other.

6.7 CNN architectures

Since 1988, when Yann LeCun (inspired by the neocognitron of Kunihiko Fukushima from 1979 (Fukushima 1980)) presented the LeNet CNN architecture to distinguish postal code numbers in images in an automatic manner (LeCun 1988), there have been a number of CNN architectures that imply more layers, different artificial network connections and distinct computational techniques in order to obtain better image detection, classification or segmentation performance (Li *et al* 2020).

Since 2010 a competition called the ImageNet Large Scale Visual Recognition Challenge (ILSVRC) (Chollet 2017) has been hosted by academics, where the main task consists in benchmarking and developing state-of-the-art algorithms to complete certain artificial vision tasks, such as object detection in images, object classification, object localization in video, etc. The ImageNet dataset is a massive annotated dataset that includes 21 841 classes and 14 197 122 images of different sizes. In 2012 the AlexNet CNN architecture won all the tasks by a relatively large margin. This was the first time CNNs were used for this task in order to achieve state-of-the-art results. Later, GoogLeNet, the VGG (developed by Stanford) and the ResNet (Lin and Jegelka 2018) architectures all presented different theoretical bases in the structure of their layers, activation functions and loss for the learning process.

ResNet (He *et al* 2015) concentrated on the fact that after using multiple sequential deep layers, the learning started to decrease performance when this theoretically should not be happening. To amend this, the loss was calculated on the residual of the loss instead of the actual loss in order to make up for these disappearing gradients in the gradient descent algorithm.

There are several popular named architectures in the field of convolutional networks. The most popular are:

1. *LeNet*: The first successful applications of convolutional networks were developed by Yann LeCun in the 1990s (LeCun *et al* 1989, Lecun *et al* 1998). Of these, the best known is the LeNet architecture that was used to read zip codes, digits, etc.
2. *AlexNet*: The first work that popularized convolutional networks was AlexNet, developed by Alex Krizhevsky, Ilya Sutskever and Geoff Hinton (Krizhevsky *et al* 2017). The network has a very similar architecture to LeNet, but is deeper and larger, and featured convolutional layers stacked on top of each other (previously it was common to only have a single convolutional layer always immediately followed by a pooling layer). This convolutional neural network was developed by researchers at the University of Toronto (later acquired by Google) and its training time viability consisted in the use of GPUs.
3. *GoogLeNet* (Inception-v4): The ILSVRC 2014 winner was a convolutional network from Szegedy *et al* from Google (Szegedy *et al* 2015). Its main contribution was the development of an *Inception module* that dramatically reduced the number of parameters in the network ($4M$, compared to AlexNet with $60M$). Additionally, they use average pooling instead of fully connected layers at the top of the convolutional network, eliminating a large amount of parameters that do not do much. The several follow-up versions are referred to as Inception versions.
4. *VGGNet*: The runner-up in ILSVRC 2014 was a network by Karen Simonyan and Andrew Zisserman that became known as VGGNet (Simonyan and Zisserman 2015). Its main contribution was in showing that the depth of the network is a critical component for good performance. Their network contained 16 convolutional/fully connected layers and features only 3×3 convolutions and 2×2 pooling from the beginning to the end. The VGGNet also is very expensive to evaluate computationally and uses more parameters than the previous networks ($138M$).
5. *ResNet*: The Residual Network developed by Kaiming He *et al* was the winner of ILSVRC 2015 (He *et al* 2015). It features special skip connections and a heavy use of batch normalization. The architecture is also missing fully connected layers at the end of the network.

   ResNets were the state-of the-art CNN model in 2016 and the basis of the architecture continues to be used in other architectures such as the ResNexts and some single shot detectors (SSDs).
6. *EfficientNet*: As of 2020 this architecture is considered to be the state-of-the-art of ImageNet using no extra training data. EfficientNets relies on the fact that it uniformly scales all dimensions of depth/width/resolution using a compound coefficient (Tan and Le 2020).

Figure 6.18. A neural network architecture. Credit: Highsmith, Carol M., 1946 - United States Library of Congress's Prints and Photographs Division under the digital ID highsm.09920.

7. *SSDs*: SSDs (Wei *et al* 2016) such as MobileNetV2 (Sandler *et al* 2019) and YOLOv4 (Redmon *et al* 2016) are small models which are leveraged for their small parameter size and are able to be deployed with very high frames-per-second (FPS) but a lower accuracy than slower models. These models usually rely on DarkNets or ResNets as a base architecture framework (figure 6.18).

Recently (as of mid-2021) Visual Transformers (ViT) closed the gap with the state-of-the-art of ImageNet, achieving 86.5% top-1 accuracy on Imagenet when training with no external data (Touvron *et al* 2021). This compares to multiple variants of the EfficientNet, which can reach up to 85.8% top 1 accuracy, training on the same ImageNet dataset only.

We hope to see more advances in the next few years concerning ViTs and will see if the current CNN paradigm for image classification may be changed.

6.8 Transfer learning

Suppose that the weights at the beginning of training are not initialized in a pseudo-random manner, but are initiated as the final state of another training task which does not necessarily have the same architecture. In this manner one can reuse the weight distribution from one classification task as a feature extractor and it will optimize more quickly or accurately in order to successfully obtain a minimum in the gradient descent task while training. This is due to the fact that similar classification tasks will be performed although the dataset images may be completely different. Some features such as colour, textures, edges, etc, are learned more rapidly.

An example in the code of listing 6.6 may be seen by reusing the ImageNet weights to classify a dataset of just cats and dogs using the ResNet-50 architecture. We see that fewer iterations are used to obtain a minimum than by just randomly

```
from __future__ import print_function

from time import time
import keras
from keras.datasets import mnist,cifar10
from keras.models import Sequential
from keras.layers import Dense, Dropout, Activation,
    Flatten
from keras.layers import Conv2D, MaxPooling2D
from keras.optimizers import Adam
from keras import backend as K

import matplotlib.pyplot as plt
import random

"""Design the CNN architecture
 number of convolutional filters to use"""
filters = 64
""" size of pooling area for max pooling"""
pool_size = 2
""" convolution kernel size"""
kernel_size = 3

"""
Extract the CIFAR data set and divide it into two groups of
    5 class data sets
"""

(x_cifar_train, y_cifar_train), (x_cifar_test, y_cifar_test
    ) = cifar10.load_data()

y_cifar_train = y_cifar_train.reshape(50000,)
y_cifar_test = y_cifar_test.reshape(10000,)

x_train_lt5 = x_cifar_train[y_cifar_train < 5]
y_train_lt5 = y_cifar_train[y_cifar_train < 5]
x_test_lt5 = x_cifar_test[y_cifar_test < 5]
y_test_lt5 = y_cifar_test[y_cifar_test < 5]

x_train_gte5 = x_cifar_train[y_cifar_train >= 5]
y_train_gte5 = y_cifar_train[y_cifar_train >= 5] - 5
x_test_gte5 = x_cifar_test[y_cifar_test >= 5]
y_test_gte5 = y_cifar_test[y_cifar_test >= 5] - 5

"""The following code will show images from the x_train_lt5
    dataset,
from the classes: airplane, automobile, bird, cat, and deer
    ."""
fig, ax = plt.subplots(2,10,figsize=(10,2.8))
```

Listing 6.6 Implementation of `train_cnn_transfer.py`.

```
fig.suptitle("Example␣of␣training␣images␣(from␣first␣5␣categories),␣for␣the␣first␣neural␣net\n", fontsize=15)
axes = ax.ravel()
for i in range(20):
    # Pick a random number
    idx=random.randint(1,1000)
    axes[i].imshow(x_train_lt5[idx])
    axes[i].axis('off')
fig.tight_layout(pad=0.5)
plt.show()

"""The following code will show images from the x_train_gte5 dataset,
from the classes: dog, frog, horse, sheep, and truck."""

fig, ax = plt.subplots(2,10,figsize=(10,2.8))
fig.suptitle("Example␣of␣training␣images␣(from␣last␣5␣categories),␣\
␣␣␣␣for␣the␣second␣neural␣net\n", fontsize=15)
axes = ax.ravel()
for i in range(20):
    # Pick a random number
    idx=random.randint(1,1000)
    axes[i].imshow(x_train_gte5[idx])
    axes[i].axis('off')
fig.tight_layout(pad=0.5)
plt.show()

num_classes = 5
input_shape = (32,32,3)
""" we define the feature layers"""
feature_layers = [
    Conv2D(filters, kernel_size,
           padding='valid',
           input_shape=input_shape),
    Activation('relu'),
    Conv2D(filters, kernel_size),
    Activation('relu'),
    MaxPooling2D(pool_size=pool_size),
    Dropout(0.25),
    Flatten(),
]
"""
we define the classification layers"""
classification_layers = [
    Dense(128),
    Activation("relu"),
    Dropout(0.25),
    Dense(num_classes),
```

Listing 6.6 (Continued.)

```
    Activation("softmax")
]
"""Our first model consists of both of the previous models sequentially added"""
model_1 = Sequential(feature_layers + classification_layers)
"""we define the training method
(here we chose Adam optimizer for quicker convergence purposes)"""
def train_model(model, train, test, num_classes):
    x_train = train[0].reshape((train[0].shape[0],) + input_shape)
    x_test = test[0].reshape((test[0].shape[0],) + input_shape)
    x_train = x_train.astype('float32')
    x_test = x_test.astype('float32')
    x_train /= 255
    x_test /= 255
    print('x_train shape:', x_train.shape)
    print(x_train.shape[0], 'train samples')
    print(x_test.shape[0], 'test samples')

    # convert class vectors to binary class matrices
    y_train = keras.utils.to_categorical(train[1], num_classes)
    y_test = keras.utils.to_categorical(test[1], num_classes)

    model.compile(loss="categorical_crossentropy",
                  optimizer=Adam(lr=0.001),
                  metrics=["accuracy"])

    t1 = time()
    model.fit(x_train, y_train,
              batch_size=batch_size,
              epochs=epochs,
              verbose=1,
              validation_data=(x_test, y_test))
    t2 = time()
    t_delta = round(t2-t1,2)
    print("Training time: {} seconds".format(t_delta))
    score = model.evaluate(x_test, y_test, verbose=0)
    print("Test score: ", score[0])
    print("Test accuracy: ", score[1])

"""
we define the number of images
entering the network at a time and the number of epochs"""

batch_size = 256
```

Listing 6.6 (Continued.)

```
epochs = 20
"""
we train the first simple model"""
train_model(model_1,
            (x_train_lt5, y_train_lt5),
            (x_test_lt5, y_test_lt5), num_classes)

"""And we visualize our training results"""
plt.title("Validation accuracy over epochs",fontsize=15)
plt.plot(model_1.history.history["val_accuracy"],lw=3,c="k")
plt.grid(True)
plt.xlabel("Epochs",fontsize=14)
plt.ylabel("Accuracy",fontsize=14)
plt.xticks([2*i for i in range(11)],fontsize=14)
plt.yticks(fontsize=14)
plt.show()


"""This freezing of feature layers is at the heart of transfer learning.
 This allows re-use of pre-trained model for classification tasks,
 because users can just stack up new fully-connected layers on top of the pre-trained feature layers
 and get good performance.

We create a fresh new model called model_2 with the untrainable feature_layers and trainable classification_layers.
 We then freeze the feature layers
"""
for l in feature_layers:
    l.trainable = False
"""
Create the second neural network model and visualize it"""

model_2 = Sequential(feature_layers + classification_layers)
model_2.summary()

"""
and then we train the model"""
# transfer: train dense layers for new classification task
train_model(model_2,
            (x_train_gte5, y_train_gte5),
            (x_test_gte5, y_test_gte5), num_classes)

"""
```

Listing 6.6 (Continued.)

```
We may now see that the accuracy seen is higher than the
    previous model
"""
plt.title("Validation␣accuracy␣over␣epochs",fontsize=15)
plt.plot(model_2.history.history["val_accuracy"],lw=3,c="k"
    )
plt.grid(True)
plt.xlabel("Epochs",fontsize=14)
plt.ylabel("Accuracy",fontsize=14)
plt.xticks([2*i for i in range(11)],fontsize=14)
plt.yticks(fontsize=14)
plt.show()
"""
And we may also do a naive comparison of the training
    duration of both models."""
plt.figure(figsize=(5,3))
plt.barh(y=["Model_1","Model_2"],width=[87,52],height=0.7)
plt.xlabel("Time␣taken␣for␣training␣(seconds)",fontsize=14)
plt.yticks(fontsize=14)
plt.xticks(fontsize=14)
plt.show()
```

Listing 6.6 (Continued.)

initializing the filters. In general, there are two types of transfer learning (Pan and Yang 2010) when applied to deep learning for computer vision:

1. Treating networks as arbitrary feature extractors.
2. Removing the fully connected layers of an existing network, placing a new FC layer set on top of the CNN and fine-tuning these weights (and optionally previous layers) to recognize object classes.

Deep neural networks trained on large-scale datasets such as ImageNet have been demonstrated to be excellent at the task of transfer learning. These networks learn a set of rich, discriminating features to recognize 1000 separate object classes. It makes sense that these filters can be reused for classification tasks other than that which the CNN was originally trained on.

Given our feature vectors, we can train an off-the-shelf machine learning model such as the linear SVM, logistic regression classifier or random forest on top of these features to obtain a classifier that recognizes new classes of images.

A mathematical definition of transfer learning may be seen as follows.

A domain D is defined as a two-element tuple consisting of feature space χ and marginal probability $P(\mathbf{X})$ where $\mathbf{X}$ is a sample data point. Thus we can represent the domain as

$$D = \mathbf{X},\ P(\mathbf{X}) \ \text{ where } \mathbf{X} = \{x_2,\ x_2,\ x_3,\ \dots,\ \alpha_n\}\ x_i \in \chi. \tag{6.35}$$

Figure 6.19. `train_cnn_transfer.py` CIFAR samples.

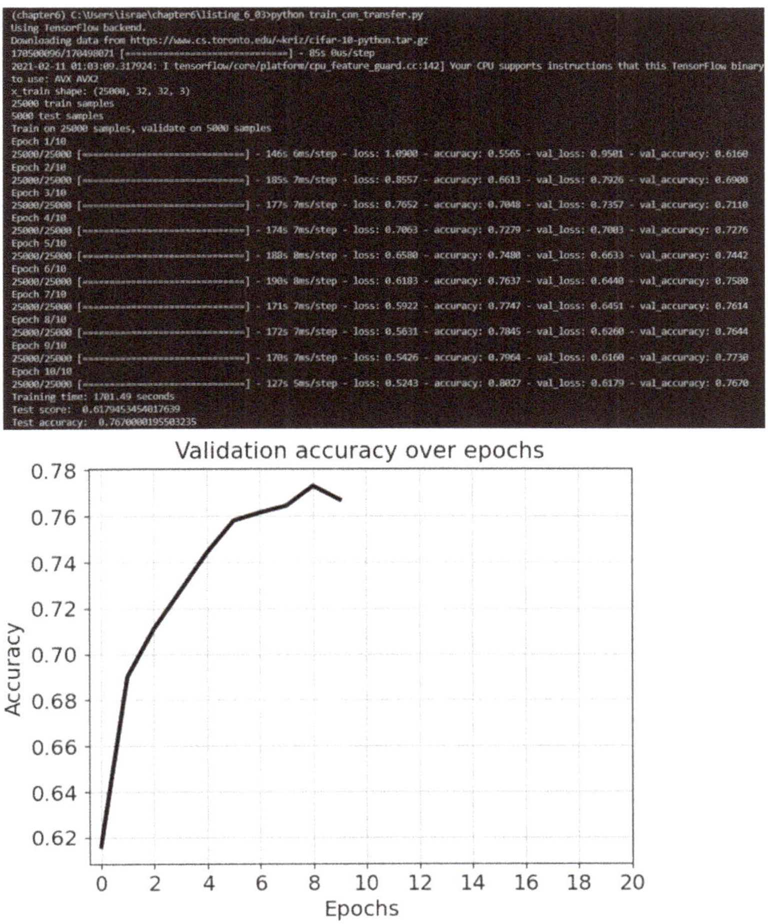

Figure 6.20. Output for listing 6.6: `train_cnn_transfer.py` epoch Model_1.

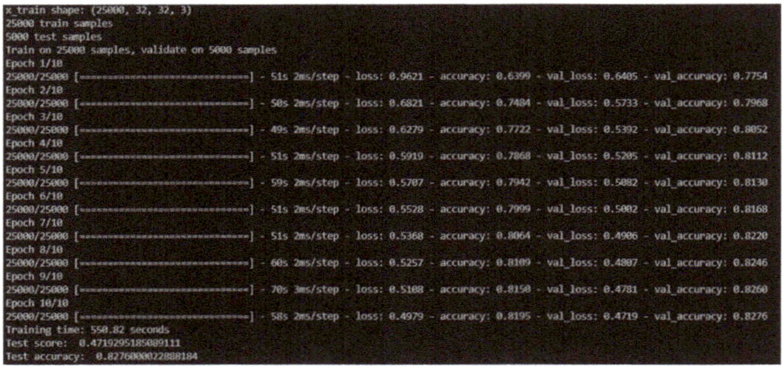

```
x_train shape: (25000, 32, 32, 3)
25000 train samples
5000 test samples
Train on 25000 samples, validate on 5000 samples
Epoch 1/10
25000/25000 [==============================] - 51s 2ms/step - loss: 0.9621 - accuracy: 0.6399 - val_loss: 0.6405 - val_accuracy: 0.7754
Epoch 2/10
25000/25000 [==============================] - 50s 2ms/step - loss: 0.6821 - accuracy: 0.7484 - val_loss: 0.5733 - val_accuracy: 0.7968
Epoch 3/10
25000/25000 [==============================] - 49s 2ms/step - loss: 0.6279 - accuracy: 0.7722 - val_loss: 0.5392 - val_accuracy: 0.8052
Epoch 4/10
25000/25000 [==============================] - 51s 2ms/step - loss: 0.5919 - accuracy: 0.7868 - val_loss: 0.5205 - val_accuracy: 0.8112
Epoch 5/10
25000/25000 [==============================] - 59s 2ms/step - loss: 0.5707 - accuracy: 0.7942 - val_loss: 0.5082 - val_accuracy: 0.8130
Epoch 6/10
25000/25000 [==============================] - 51s 2ms/step - loss: 0.5528 - accuracy: 0.7999 - val_loss: 0.5002 - val_accuracy: 0.8168
Epoch 7/10
25000/25000 [==============================] - 51s 2ms/step - loss: 0.5368 - accuracy: 0.8064 - val_loss: 0.4906 - val_accuracy: 0.8220
Epoch 8/10
25000/25000 [==============================] - 60s 2ms/step - loss: 0.5257 - accuracy: 0.8109 - val_loss: 0.4807 - val_accuracy: 0.8246
Epoch 9/10
25000/25000 [==============================] - 70s 3ms/step - loss: 0.5108 - accuracy: 0.8150 - val_loss: 0.4781 - val_accuracy: 0.8260
Epoch 10/10
25000/25000 [==============================] - 58s 2ms/step - loss: 0.4979 - accuracy: 0.8195 - val_loss: 0.4719 - val_accuracy: 0.8276
Training time: 550.82 seconds
Test score:  0.4719295185089111
Test accuracy:  0.8276000022888184
```

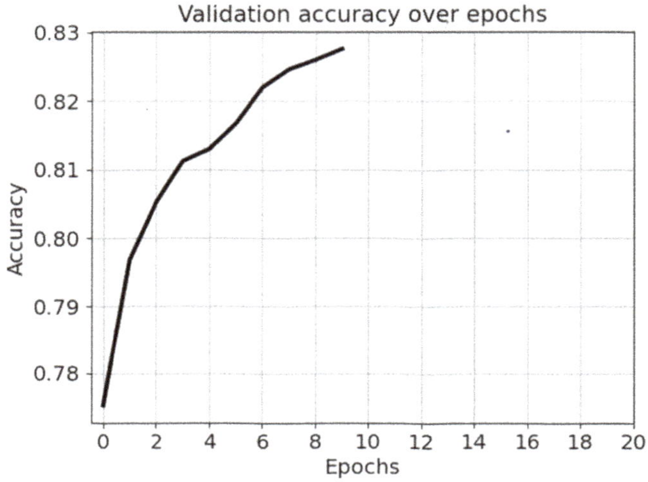

Figure 6.21. Output for listing 6.6: `train_cnn_transfer.py` epoch Model_2.

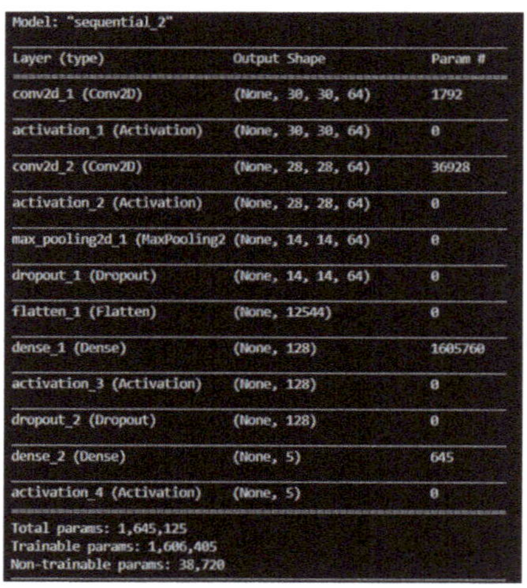

```
Model: "sequential_2"
_________________________________________________________________
Layer (type)                 Output Shape              Param #
=================================================================
conv2d_1 (Conv2D)            (None, 30, 30, 64)        1792
_________________________________________________________________
activation_1 (Activation)    (None, 30, 30, 64)        0
_________________________________________________________________
conv2d_2 (Conv2D)            (None, 28, 28, 64)        36928
_________________________________________________________________
activation_2 (Activation)    (None, 28, 28, 64)        0
_________________________________________________________________
max_pooling2d_1 (MaxPooling2 (None, 14, 14, 64)        0
_________________________________________________________________
dropout_1 (Dropout)          (None, 14, 14, 64)        0
_________________________________________________________________
flatten_1 (Flatten)          (None, 12544)             0
_________________________________________________________________
dense_1 (Dense)              (None, 128)               1605760
_________________________________________________________________
activation_3 (Activation)    (None, 128)               0
_________________________________________________________________
dropout_2 (Dropout)          (None, 128)               0
_________________________________________________________________
dense_2 (Dense)              (None, 5)                 645
_________________________________________________________________
activation_4 (Activation)    (None, 5)                 0
=================================================================
Total params: 1,645,125
Trainable params: 1,606,405
Non-trainable params: 38,720
_________________________________________________________________
```

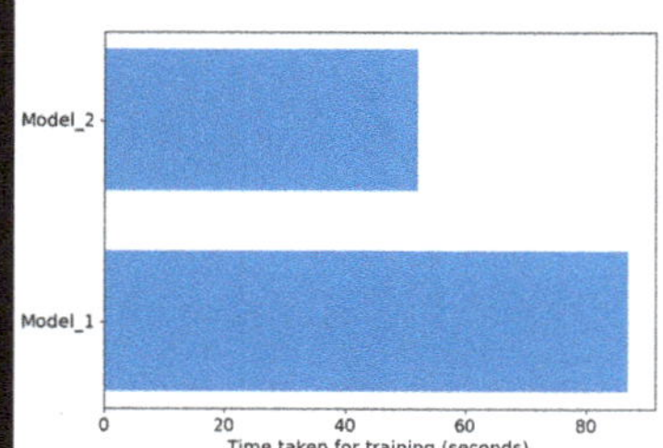

Figure 6.22. Output for listing 6.6: `train_cnn_transfer.py` comparison between models.

For a given domain D a task T is defined by two components: a label space Y and a predictive function γ learned from the feature vector/label pair x_i in X, y_i in Y. Each feature vector from the domain D, η predicts the corresponding label $\eta(x_i) = y_i$.

Then, given a source domain D_s, a corresponding source task T_s and a target domain D_t as well as a task target T_t, the objective of transfer learning is to enable the learning of the target conditional probability distribution $P(Y_t|X_t)$ in D_t with the information gained from D_s and T_s, where $D_s \neq D_t$ or $T_s \neq T_t$.

Next, we will use the CIFAR-10 dataset to illustrate a use of transfer learning (see listing 6.6 for the code and figures 6.19–6.22 for the samples and output).

6.9 End notes

Deep learning currently dominates most vision activities but this does not make deep learning models necessarily the best option from deployment, to training; the least resource dependent is classical vision which still plays a big part in solving many vision tasks. If the level of abstraction becomes higher and the available data (that is in the order of tens of thousands) may be labeled as clean as possible having deployment conditions similar to the training settings, then this will always be the most accurate and ideal way to go. Newer vision architectures such as visual-transformers seem to be catching up in performance but the heavy resource dependency in the training will not prove to be as easy to implement as CNNs.

References

Arora R, Basu A, Mianjy P and Mukherjee A 2018 Understanding deep neural networks with rectified linear units ICLR 2018 Conference Blind Submission arXiv: 1611.01491

Cesa-Bianchi N, Long P M and Warmuth M K 1996 Worst-case quadratic loss bounds for prediction using linear functions and gradient descent *IEEE Trans. Neur. Netw.* **7** 604–19

Chollet F 2017 Xception: deep learning with depthwise separable convolutions *2017 IEEE Conf. on Computer Vision and Pattern Recognition (CVPR)* pp 1800–7

Fukushima K 1980 Neocognitron: A self-organizing neural network model for a mechanism of pattern recognition unaffected by shift in position *Biol. Cybernet.* **36** 193–202

Glorot X, Bordes A and Bengio Y 2011 Deep sparse rectifier neural networks *JMLR Workshop and Conf. Proc.* vol 15 pp 315–23

He J, Li L, Xu J and Zheng C 2020 ReLU deep neural networks and linear finite elements *J. Comput. Math.* **38** 502–27

He K, Zhang X, Ren S and Sun J 2015 *Deep Residual Learning for Image Recognition* arXiv: 1512.03385

Kim Y 2014 *Convolutional Neural Networks for Sentence Classification* (Association for Computational Linguistics) pp 1746–51 arXiv: 1408.5882

Krizhevsky A, Sutskever I and Hinton G E 2017 ImageNet classification with deep convolutional neural *Commun. ACM* **60** 84–90

LeCun Y 1988 A theoretical framework for back-propagation *Proc. of the 1988 Connectionist Models Summer School, CMU (Pittsburgh, PA)* D Touretzky, G Hinton and T Sejnowski (San Mateo, CA: Morgan Kaufmann) pp 21–8

LeCun Y, Jackel L D, Boser B, Denker J S, Graf H P, Guyon I, Henderson D, Howard R E and Hubbard W 1989 *Handwritten Digit Recognition: Applications of Neural Net Chips and Automatic Learning* (New York: IEEE Communication) pp 41–6

Lecun Y, Bottou L, Bengio Y and Haffner P 1998 Gradient-based learning applied to document recognition *Proc. IEEE* **86** 2278–24
Li Z, Yang W, Peng S and Liu F 2020 *A Survey of Convolutional Neural Networks: Analysis, Applications, and Prospects* arXiv: 2004.02806
Lin H and Jegelka S 2018 *ResNet with One-Neuron Hidden Layers is a Universal Approximator* arXiv: 1806.10909
Mamalet F and Garcia C 2012 Simplifying CovNets for fast learning *Artificial Neural Networks and Machine Learning—ICANN 2012* (Berlin: Springer) pp 58–65
Minsky M L and Papert S A 1969 *Perceptrons* (Cambridge, MA: PIT Press)
Minsky M L and Papert S A 1988 *Perceptrons (expanded edn)* (Cambridge, MA: MIT Press)
Mitchell T, Buchanan B, DeJong G, Dietterich T, Rosenbloom P and Waibel A 1990 *Machine learning Annu. Rev. Comput. Sci. B* **4** 417–33
Mitchell T, Buchanan B, DeJong G, Dietterich T, Rosenbloom P and Waibel A 1990 Machine learning *Annu. Rev. Comput. Sci.* B **4** 417–33
O'Shea K and Nash R 2015 *An Introduction to Convolutional Neural Networks* arXiv: 511.08458
Pan S J and Yang Q 2010 A survey on transfer learning *IEEE Trans. Knowl. Data Eng.* **22** 1345–59
Pitts W and McCulloch W S 1943 A logical calculus of the ideas immanent in nervous activity *Bull. Math. Biophys.* **5** 115–33
Pope P E, Kolouri S, Rostami M, Martin C E and Hoffmann H 2019 Explainability methods for graph convolutional neural networks *2019 IEEE/CVF Conf. on Computer Vision and Pattern Recognition (CVPR)* 10764–73
Redmon J, Divvala S, Girshick R and Farhadi A 2016 You only look once: unified, real-time object detection *Proc. of the IEEE Conf. on Computer Vision and Pattern Recognition (CVPR)*
Rojas R 1996 *Neural Networks: A Systematic Introduction* (Berlin: Springer)
Rosenblatt F 1958 The perceptron: a probabilistic model for information storage and organization in the brain *Psychol. Rev.* **65** 386–408
Sandler M, Howard A, Zhu M, Zhmoginov A and Chen L-C 2019 MobileNetV2: Inverted residuals and linear bottlenecks arXiv: 1801.04381
Simonyan K and Zisserman A 2015 Very deep convolutional networks for large-scale image recognition arXiv: 1409.1556
Szegedy C, Wei L, Yangqing J, Sermanet P, Reed S, Anguelov D, Erhan D, Vanhoucke V and Rabinovich A 2015 Going deeper with convolutions *2015 IEEE Conf. on Computer Vision and Pattern Recognition (CVPR)* pp 1–9
Tan M and Le Q V 2020 EfficientNet: rethinking model scaling for convolutional neural networks arXiv: 1905.11946
Touvron H, Cord M, Sablayrolles A, Synnaeve G and Jégou H 2021 Going deeper with image transformers arXiv: 2103.17239
Wei L, Dragomir A, Dumitru E, Christian S, Scott R, Cheng-Yang F and Berg A C 2016 *SSD: Single Shot MultiBox Detector* (Lecture Notes in Computer Science) (Berlin: Springer) pp 21–37
Yehudai G and Shamir O 2020 On the power and limitations of random features for understanding neural networks arXiv: 1904.00687
Zeiler M D and Fergus R 2014 Visualizing and understanding convolutional networks *Computer Vision—ECCV 2014 (New York)* pp 818–33

IOP Publishing

Optics and Artificial Vision

Rafael G González-Acuña, Héctor A Chaparro-Romo and Israel Melendez-Montoya

Chapter 7

Optical character recognition

In this chapter we explore the concepts behind optical character recognition (OCR) and the main problems and applications that can be solved by using OCR. Then we focus on a simple implementation of OCR in `Python` (van Rossum 1995) with `pytesseract`. We explain step-by-step a code that computes OCR over an image without `pytesseract`. Finally we discuss the limitations of `pytesseract`.

7.1 Introduction

OCR is a process aimed at the digitization of texts, which automatically identifies symbols or characters that belong to a certain alphabet from an image and then stores them as data. The end product is to extract this text and to address this task there exist two primary branches of OCR: document OCR and text recognition in the wild. This second task is a much more difficult problem and various solutions exist up to a certain accuracy. This is one of the reasons why we cannot say that OCR is a *solved* problem. As in the artificial tasks that perform facial recognition, the main problems with text detection in the wild occur when different orientations, lighting, noise, background objects and fonts are present.

OCR had long been a problem with different solutions before Yann LeCun automatized a manner of reading postal code numbers with neural networks in 1989 (LeCun *et al* 1989). The OCR problem varies a lot due to the many factors that may contribute in the context of obtaining the image to process. In most usual contexts OCR problems can be solved using several morphological transformation techniques (Gao *et al* 2010, Gonzalez *et al* 2004, Caulfield and Maloney 1969, Cash and Hatamian 1987), although state-of-the-art OCR uses deep learning techniques (Islam *et al* 2017). OCR tools inherently lack the intelligence to parse or understand extracted text beyond just extracting it. Thus we will not get into the subject of natural language processing (NLP), which may be frequently coupled with these tools.

In this chapter we will review the general conceptual manner of solving an OCR problem without having to resort to neural networks (NNs) (Charles *et al* 2012) and we will also briefly review the basis that is used to train deep learning architectures to solve OCR problems. One can also use other deep learning techniques, such as recurrent convolutional neural networks (RCNNs), visual transformers (ViTs) and generative adversarial networks (GANs), with CNNs to solve OCR (Kapur 2017), but the authors will limit the contents of this chapter to the conceptual basics and applications of OCR.

There still exist several challenges in fine-tuning OCR techniques, some as complicated as using OCR in polygon-like word detection (Rao *et al* 2016, Chaudhuri *et al* 2017. One may see several of these OCR challenges in the competitions hosted by the International Conference on Document Analysis and Recognition (ICDAR), where more challenges continue to be added each year. In these competitions we may see, for example, that there exists a Robust Reading Challenge (RRC) from 2017 where different engineering techniques may be applied for text detection in the specifically obtained COCO-text dataset, there also exists a Robust Reading Challenge on Arbitrary-Shaped Text using a competition specific ArT dataset. In general, there is no one-size-fits-all model and one must usually use *fine-tuning* in order to obtain the highest classification or segmentation metrics.

7.2 Problems in classical OCR

The basic process that is carried out in OCR is to convert the text that appears in an image into a text file that can be edited and used as such by any other programme or application that needs it. Each problem in OCR is different as every application has different needs depending on the input restrictions. However, usually an OCR should first detect the text in an image and then give an output of where each letter, word and punctuation corresponds to in the image text. The usual OCR pipeline uses a text detector which localizes the text in an image and then a classifier which tells us what each individual character is. Using deep learning this is a fairly accurately performing but time intensive task which depends on the specificity of the detection task at hand (e.g. handwritten text recognition, document recognition, cursive letter recognition, text in the wild, etc) and the availability of the labelled dataset. In classical OCR we have to '*manually*' pinpoint thresholds and parameters (Forsyth and Ponce 2002) for the task at hand, but this could nonetheless be a faster and less time intensive solution than using deep learning if an already trained network for the task at hand is not readily available (Mori *et al* 1999, Nagy *et al* 1999, Mohammad *et al* 2014).

Starting from a perfect image (Granlund and Knutsson 2013, Haralick *et al* 1987, Chityala and Pudipeddi 2020), that is, an image with only two levels in greyscale, the recognition of these characters will basically be carried out by comparing them with patterns or templates thaeamt contain all the possible characters. Note that the actual images are not perfect in the sense that the same number or letter will not be *exactly* the same in the same image, therefore OCR runs into several problems:

1. The imaging device may introduce grey levels to the background that do not belong in the original image.
2. The resolution of these devices can introduce noise into the image, affecting the pixels to be processed.

3. The distance that separates some characters from others, not always being the same, can cause recognition errors.
4. Connecting two or more characters with common pixels can also cause errors.

Clearly, if part two is done well, part three is easy either with pattern matching or using machine learning classification algorithms. However, the contour detection is quite challenging in the generalization process concerning different images. Much manual fine-tuning is usually required for this, and therefore becomes infeasible in most problems.

7.3 The basic scheme of a classical OCR algorithm

All classical OCR algorithms are intended to be able to differentiate a text from any background image. To do this they are based on four 'manual' stages:

1. *Binarization or characterization*: Apply filters to make the characters stand out from the background (Vincent and Folorunso 2009).
2. *Fragmentation or segmentation of the image*: Apply contour detection to recognize the characters one by one (White and Rohrer 1983).
3. *Component thinning*: Refine each character that was found (Sreedhar 2012).
4. *Comparison with patterns*: Apply image classification to identify the characters (Xu *et al* 2011, He *et al* 2013).

7.3.1 Binarization

Most OCR algorithms are based on a binary image (two colours). Therefore it is convenient to convert a greyscale image, or a colour one, to a black and white image in such a way that the essential properties of the image are preserved (Schalkoff 1989, Umesh 2012). One way to do this is through the image histogram, which shows the number of pixels for each level of grey that appears in the image. To binarize this we have to choose an appropriate threshold, from which all the pixels that do not exceed it will become black and the rest white (Klette 1980).

Through this process we obtain a black and white image where the contours of the characters and symbols contained in the image are clearly marked. From here we can isolate the parts of the image that contain text (more transitions between black and white).

7.3.2 Fragmentation or segmentation of the image

This is the most complex, variant and necessary process for subsequent character recognition. Image segmentation involves detection by means of 'deterministic labelling' or stochastic procedures of the contours or regions of the image, based on intensity information or spatial information. It allows the decomposition of a text into different logical entities, which must be sufficiently invariable to be independent of the writer and sufficiently significant for its recognition. There is no generic method to carry out this image segmentation that is efficient enough for the analysis of a text. Although the most commonly used techniques are variations of the methods based on linear projections. One of the most classic and simple techniques for grey-level images consists of determining the modes or groupings (clusters) from the histogram, in such a way that they allow a classification or thresholding of the pixels in homogeneous regions.

7.3.3 Component thinning

Once the related components of the image have been isolated, a thinning process will have to be applied to each of them. This procedure consists of successively erasing the points of the contours of each component so that its typology is preserved. The elimination of the points must follow a scheme of successive sweeps so that the image continues to have the same proportions as the original and thus ensure that it is not deformed. You have to sweep in parallel, that is, mark the erasable pixels to eliminate them all at once. This procedure is carried out to make classification and recognition possible, simplifying the shape of the components.

7.3.4 Comparison with patterns

At this stage, the characters obtained previously are compared with theoretical ones (patterns) stored in a database. The proper functioning of the OCR is largely based on a good definition of this stage. There are different methods to carry out the comparison. One of them is the projection method, in which vertical and horizontal projections of the character to be recognized are obtained, and they are compared with the alphabet of possible characters (of a certain font) until the maximum match is found.

7.4 Classical OCR using machine learning

We now present the basis of classical OCR using a machine learning *K*-nearest neighbours (KNN) algorithm (Klette 2014, Raschka 2015) to distinguish between numbers. This example is taken from a StackOverflow answer originating from the question '*Simple digit recognition OCR in* `OpenCV-Python`' which was responded to by user Abid Rahman K. The task at hand is to be able to distinguish the individual numbers of Pi in a plain binary format using a set of training `samples`. Figure 7.1 shows the characters used as training data.

See listing 7.1 for the `Python` code. The first step is to load the image and apply preprocessing in order to obtain each digit separately. A Gaussian blur

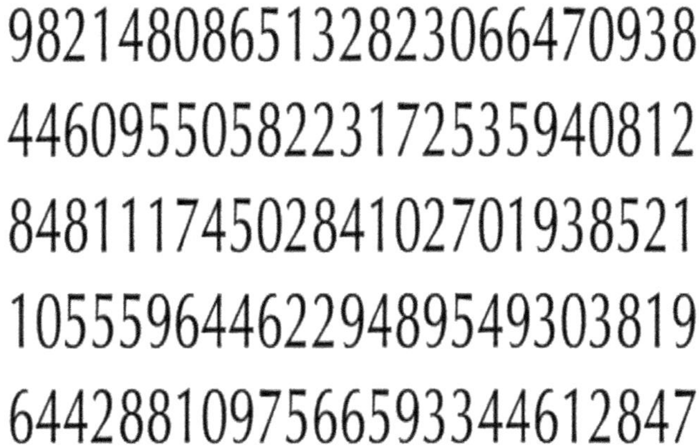

Figure 7.1. Characters used for training our OCR algorithm.

```python
import sys
import numpy as np
import cv2

im = cv2.imread("training_data.png")
im3 = im.copy()

gray = cv2.cvtColor(im,cv2.COLOR_BGR2GRAY)
blur = cv2.GaussianBlur(gray,(5,5),0)
thresh = cv2.adaptiveThreshold(blur,255,1,1,11,2)
contours, hierarchy = cv2.findContours(thresh,cv2.RETR_LIST
    ,cv2.CHAIN_APPROX_SIMPLE)
samples =  np.empty((0,100))
responses = []
keys = [i for i in range(48,58)]

for cnt in contours:
    if cv2.contourArea(cnt)>50:
        [x,y,w,h] = cv2.boundingRect(cnt)

        if  h>28:
            cv2.rectangle(im,(x,y),(x+w,y+h),(0,0,255),2)
            roi = thresh[y:y+h,x:x+w]
            roismall = cv2.resize(roi,(10,10))
            cv2.imshow("norm", im)
            key = cv2.waitKey(0)

            if key == 27:  # (escape to quit)
                sys.exit()
            elif key in keys:
                responses.append(int(chr(key)))
                sample = roismall.reshape((1,100))
                samples = np.append(samples,sample,0)

responses = np.array(responses,np.float32)
responses = responses.reshape((responses.size,1))
print("training complete")

np.savetxt("generalsamples.data",samples)
np.savetxt("generalresponses.data",responses)
```

Listing 7.1 The code for `simple_ocr_1.py`.

(Shen *et al* 2007) may be applied in order to eliminate any Gaussian noise in the image that may be caused by a low quality image. This also slightly extends the edges of our numbers for positioning our bounding boxes. We then use the `findContours` `OpenCV` function in order to obtain the contour of each digit.

Then we use `boundingRect` in order to find the rectangle coordinates of each digit. We filter the contours according to height and area to make sure we obtain only the separated digits. Each digit inside a red bounding box is then displayed and we must label it accordingly by typing the corresponding digit on our keyboard. We resize each digit as `samples` in order to have consistency in image size and therefore feature vector size to apply a kNN algorithm. We then save the training feature vectors together with their corresponding label in the text files `generalsamples.data` and generalresponses.data. The primary task is to be able to correctly identify each digit in the image called `pi_image.png` (figure 7.2). Thus we realize a training method that uses the features of each bounded digit, which in this case are the pixel values in each bounding box. See listing 7.2 for the `Python` code.

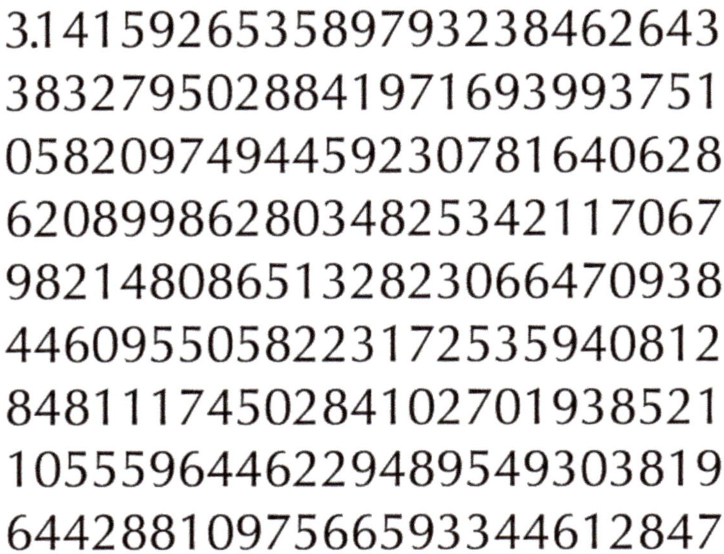

Figure 7.2. The characters used for training our OCR algorithm with a π value.

```
import cv2
import numpy as np

samples = np.loadtxt('generalsamples.data',np.float32)
responses = np.loadtxt('generalresponses.data',np.float32)
responses = responses.reshape((responses.size,1))

model = cv2.ml.KNearest_create()
model.train(samples,cv2.ml.ROW_SAMPLE,responses)

im = cv2.imread("pi_data.png")
out = np.zeros(im.shape,np.uint8)
gray = cv2.cvtColor(im,cv2.COLOR_BGR2GRAY)
thresh = cv2.adaptiveThreshold(gray,255,1,1,11,2)

contours, hierarchy = cv2.findContours(thresh,cv2.RETR_LIST
    ,cv2.CHAIN_APPROX_SIMPLE)[-2:]
```

Listing 7.2 The code for `simple_ocr_2.py`.

```
for cnt in contours:
    if cv2.contourArea(cnt)>50:
        [x,y,w,h] = cv2.boundingRect(cnt)
        if  h>28:
            cv2.rectangle(im,(x,y),(x+w,y+h),(0,255,0),2)
            roi = thresh[y:y+h,x:x+w]
            roismall = cv2.resize(roi,(10,10))
            roismall = roismall.reshape((1,100))
            roismall = np.float32(roismall)
            retval, results, neigh_resp, dists = model.
               findNearest(roismall, k = 1)
            string = str(int((results[0][0])))
            cv2.putText(out,string,(x,y+h),0,1,(0,255,0))

cv2.imshow("im",im)
cv2.imshow("out",out)
cv2.waitKey(0)
```

Listing 7.2 (Continued.)

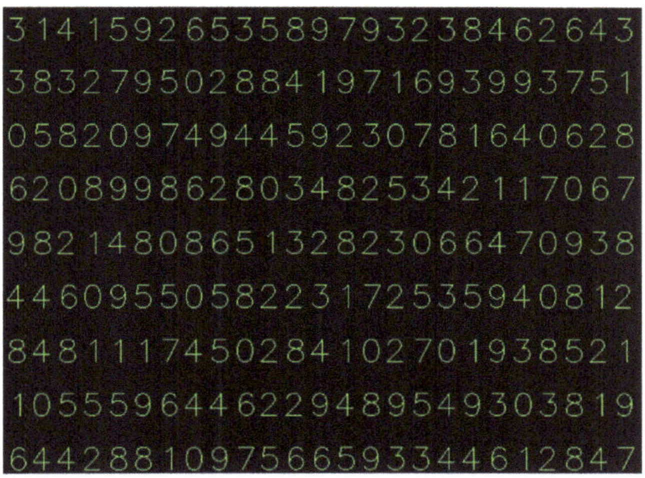

Figure 7.3. Output for listing 7.2 using the π value image.

The `KNearest_create().train` method uses the feature vectors of each training digit and calculates the Euclidean distances between the feature vector (or digit) to classify. In this case we use $K = 1$ because the feature vectors that are closest to the unclassified digit are the most alike in pixel intensity and this will be our main decision in our classification. Applying this algorithm we are indirectly performing a type of template matching. We finally visualize our results depending on how well we labelled our dataset, as can be seen in figure 7.3.

This simple OCR computer vision script gives us the basic understanding of how OCR usually works. Here we use a machine learning algorithm called kNN to sort out the difference between the ten characters. We prefer this technique to pattern matching or template matching as these other techniques are usually more brute force algorithms (Szeliski 2010).

7.5 Modern OCR with deep learning

After the 2012 boom of convolutional networks, most state-of-the-art image classification and segmentation has been based on CNNs (Mithe *et al* 2013). This game-changer also applied strongly to the field of OCR and its different problems. In Github we may see that there exists a very high number of OCR papers and open source codes that give solutions to specific OCR tasks and thus when using deep learning we must carefully choose a task to solve and select an adequate training dataset. Since the advent of OCR algorithms, there have been many services that have introduced these procedures to increase their performance and others that are completely based on these technologies. Below we discuss some of the most notable applications that use OCR.

7.5.1 Handwritten text recognition

The difficulties that we find when it comes to recognizing a typed text cannot be compared to those that appear when we want to recognize handwritten text. We do not all write uniformly and we do not all write the number 4 in the same way, for example. To address these types of problems specific techniques and applications have been developed which are called intelligent character recognition (ICR). Although text is basically made up of individual characters, most OCR algorithms do not perform well on handwriting, since segmenting continuous text is a complex procedure (Arica and Yarman-Vural 2002).

In the case of handwriting recognition in the process of exam correction, there is the possibility, by adding a list of lexicons (names and surnames), to approach 100% correctness. Through the ICR response boxes you can recognize words, such as country names, region names, trademarks, etc, in short, everything that can be integrated into a list of words (lexicon) which can be increased according to the needs.

In the real world sometimes a sentence can only be understood when we have finished reading it. Automating this procedure involves an operation of morphological, lexical and syntactic levels that is achieved through the recognition of continuous speech. To carry out this methodology robust algorithms are used that employ prior segmentation, since it is obtained automatically with decoding.

See Mancas-Thillou and Gosselin's work (Thillou and Gosselin 2007) if you are further interested in the challenges associated with text detection in natural scene images (Thillou and Gosselin 2007).

7.5.2 Indexing with databases

With the great increase in published information that has taken place in recent years, more and more methods are used to organize all this material stored in databases. One of these types of content is images. One of the most common ways to search for images is from manually entered metadata by users. Currently, search engines have appeared that provide the possibility of searching for images by means of the text that appears in them (Klette and Zamperoni 1996, López *et al* 2016), such as the Document Image Retrieval System (DIRS) search engine that, through an OCR algorithm, extracts the text that appears in the image and uses it as metadata that can be used for searches. This technology provides a possibility for image searching and shows that, despite its limitations, OCR can still achieve a lot.

7.6 OCR with Tesseract

Tesseract (Patel *et al* 2012) is an OCR engine for various operating systems. It is a free software, released under the Apache license, Version 2.02, and its development has been funded by Google since 2006. Tesseract was originally developed as proprietary software at Hewlett-Packard laboratories in Bristol (UK) and Greeley (CO, USA) between 1985 and 1994. After ten years without any development, it was released as open source in 2005 by Hewlett-Packard and the University of Nevada, Las Vegas.

There is a `Python` (Van Rossum and Drake 1995) version of Tesseract called `pytesseract`. Here we are going to implement it to obtain strings from images. `pytesseract` works as follows. First it applies an adaptive thresholding which converts the image into a binary image. Then the next step is to compute the connected component analysis, which labels all the resulting elements of the binary image. Connected component analysis is used to extract character outlines. This step is particularly useful because it performs the OCR of an image with white text and a black background. Then after the mentioned steps, `pytesseract` seeks to recognize text lines. Text lines are analysed for some predefined area or equivalent text size. Then the text is divided into words using fuzzy spaces.

Each word passed as satisfactory is passed on to an adaptive classifier as training data. The adaptive classifier tries to recognize text in a more accurate manner. The output of the process implemented in `pytesseract` is presented in figure 7.4. The `Python` code in listing 7.3 obtains a string from a given image. This code implements `pytesseract`. For the use of this code `pytesseract` should be satisfactorily installed.

The following explains the code step-by-step:

1. The first lines of the code in listing 7.3 import the necessary packages to perform OCR over an image. Then we construct the argument parse, as we have done in our codes in the previous chapters.
2. Then we load the image to which we are going to apply the OCR under the name `image` with the argument `-image`. Then we call `grey` for the conversion of `image` to greyscale.
3. The next section of the code applies a threshold or a blurring filter over `grey` if it is required. Once the threshold and filter steps are performed we write `grey` on a temporary file called `filename`.

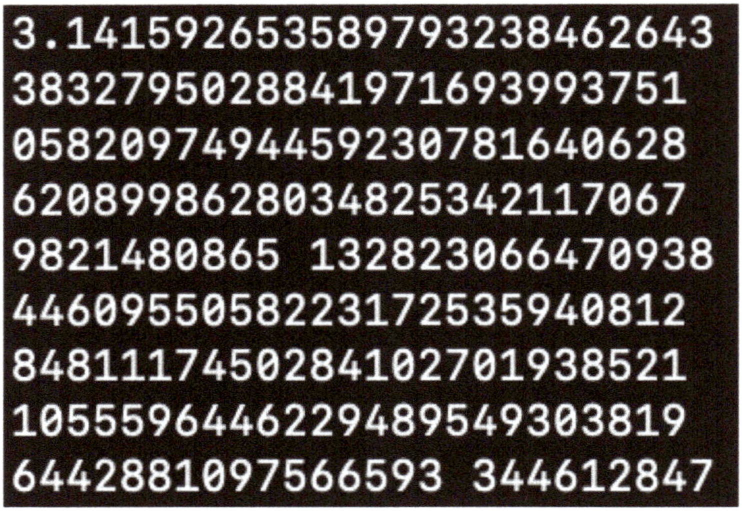

Figure 7.4. Terminal output of `OCR.py`.

```
# load  packages
from PIL import Image
import pytesseract
import argparse
import cv2
import os

# construct the argument parse and parse the arguments
ap = argparse.ArgumentParser()
ap.add_argument("-i", "--image", required=True,
        help="path to input image to be OCR'd",
        default ="pi_data.png")
ap.add_argument("-p", "--preprocess", type=str, default="thresh",
        help="type of preprocessing to be done")
args = vars(ap.parse_args())

# load image grayscale
image = cv2.imread(args["image"])
# convert image to grayscale
gray = cv2.cvtColor(image, cv2.COLOR_BGR2GRAY)

# check  if the thresh is needed
if args["preprocess"] == "thresh":
        gray = cv2.threshold(gray, 0, 255,
                cv2.THRESH_BINARY | cv2.THRESH_OTSU)[1]
# check  if the blurring is need it
elif args["preprocess"] == "blur":
```

Listing 7.3 `OCR.py`.

```
        gray = cv2.medianBlur(gray, 3)

# name a temporary file
filename = "{}.png".format(os.getpid())
# write the grayscale image to disk as a temporary file
cv2.imwrite(filename, gray)

# load the image as a PIL/Pillow image, apply OCR, and then
text = pytesseract.image_to_string(Image.open(filename))
# delete the temporary file
os.remove(filename)
# print the text of the image
print(text)

# show the output images
cv2.imshow("Image", image)
cv2.imshow("Output", gray)
cv2.waitKey(0)
```

Listing 7.3 (Continued.)

4. With the command `Image.open(filename)` we open `filename` and we apply the OCR over it with the command `pytesseract.image_to_string`. This last command give us a string that we call `text`.
5. Finally we print the string `text`.

It is very important to mention that `pytesseract.image_to_string` can incorrectly distinguish words or characters if the image has a lot of noise. It is a matter of preprocessing it to fragmented texts using the command implemented in the algorithm above, `pytesseract.image_to_string`.

Figure 7.4 is the output of putting in `pi_data.png`. In order to test `pytesseract` we apply `pytesseract.image_to_string` over them and we display the results. Note that `pytesseract` may fail in some instances even if the images have excellent quality due to rotations, spacing, deformations, noise masks, etc.

7.7 End notes

In this chapter we gave a general description of OCR and we studied the problems that OCR addresses. We were able to recognize that much of the preprocessing that is used in OCR uses algorithms already known and studied in previous chapters. Finally, we present an OCR and `Python` code, study its performance and discuss its limitations.

References

Arica N and Yarman-Vural F T 2002 Optical character recognition for cursive handwriting *IEEE Trans. Pattern Anal. Mach. Intell.* **24** 801–13

Cash G L and Hatamian M 1987 Optical character recognition by the method of moments *Comput. Vis. Graph. Image Process.* **39** 291–310

Caulfield H J and Maloney W T 1969 Improved discrimination in optical character recognition *Appl. Opt.* **8** 2354–6

Charles P K, Harish V, Swathi M and Deepthi C H 2012 A review on the various techniques used for optical character recognition *Int. J. Eng. Res. Appl.* **2** 659–62

Chaudhuri A, Mandaviya K, Badelia P and Ghosh S K 2017 Optical character recognition systems *Optical Character Recognition Systems for Different Languages with Soft Computing* (Berlin: Springer) pp 9–41

Chityala R and Pudipeddi S 2020 *Image Processing and Acquisition Using* `Python` (Boca Raton, FL: CRC Press)

Forsyth D A and Ponce J 2002 *Computer Vision: A Modern Approach* (Englewood Cliffs, NJ: Prentice Hall)

Gao W, Zhang X, Yang L and Liu H 2010 An improved Sobel edge detection *2010 3rd Int. Conf. on Computer Science and Information Technology* (Piscataway, NJ: IEEE) pp 67–71

Gonzalez R C, Woods R E and Eddins S L 2004 *Digital Image Processing Using MATLAB* (New Delhi: Pearson Education India)

Granlund G H and Knutsson H 2013 *Signal Processing for Computer Vision* (Berlin: Springer)

Haralick R M, Sternberg S R and Zhuang X 1987 Image analysis using mathematical morphology *IEEE Trans. Pattern Anal. Mach. Intell.* **PAMI-9** 532–50

He Z-Y, Sun L-N and Chen L-G 2013 Fast computation of threshold based on Otsu criterion *Dianzi Xuebao (Acta Electron. Sin.)* **41** 267–72

Islam N, Islam Z and Noor N 2017 A survey on optical character recognition system arXiv:1710.05703

Kapur S 2017 *Computer Vision with* `Python` *3* (Birmingham: Packt)

Klette R 1980 Parallel operations on binary images *Comput. Graph. Image Process* **14** 145–58

Klette R 2014 *Concise Computer Vision* (Berlin: Springer)

Klette R and Zamperoni P 1996 *Handbook of Image Processing Operators* (Hoboken, NJ: Wiley)

LeCun Y, Jackel L D, Boser B, Denker J S, Graf H P, Guyon I, Henderson D, Howard R E and Hubbard W 1989 *Handwritten Digit Recognition: Applications of Neural Net Chips and Automatic Learning* (New York: IEEE Communication) pp 41–6

López A F J, Pelayo M C P and Forero Á R 2016 Teaching image processing in engineering using `Python` *IEEE Rev. Iberoam. Tecnol. Aprendizaje.* **11** 129–36

Mithe R, Indalkar S and Divekar N 2013 Optical character recognition *Int. J. Recent Technol. Eng.* **2** 72–5

Mohammad F, Anarase J, Shingote M and Ghanwat P 2014 Optical character recognition implementation using pattern matching *Int. J. Comput. Sci. Inform. Technol.* **5** 2088–90

Mori S, Nishida H and Yamada H 1999 *Optical Character Recognition* (New York: Wiley)

Nagy G, Nartker T A and Rice S V 1999 Optical character recognition: an illustrated guide to the frontier *Document Recognition and Retrieval VII* vol 3967 (Bellingham, WA: International Society for Optics and Photonics) pp 58–69

Patel C, Patel A and Patel D 2012 Optical character recognition by open source OCR tool tesseract: a case study *Int. J. Comput. Appl.* **55** 50–6

Rao N V 2016 Optical character recognition technique algorithms *Theor. Appl. Inform Technol.* **83** 275–82

Raschka S 2015 *Python Machine Learning* (Birmingham: Packt)

Schalkoff R J 1989 *Digital Image Processing and Computer Vision* vol 286 (New York: Wiley)

Shen B, Xu Y, Lu G and Zhang D 2007 Detecting iris lacunae based on Gaussian filter *Third Int. Conf. on Intelligent Information Hiding and Multimedia Signal Processing (IIH-MSP 2007)* vol 1 (Piscataway, NJ: IEEE) pp 233–6

Sreedhar K 2012 Enhancement of images using morphological transformations *Int. J. Comput. Sci. Inform. Technol.* **4** 33–50

Szeliski R 2010 *Computer Vision: Algorithms and Applications* (Berlin: Springer)

Thillou C M and Gosselin B 2007 Natural scene text understanding *Vision Systems: Segmentation and Pattern Recognition* ed G Obinata and A Dutta (Rijeka: IntechOpen))

Umesh P 2012 Image processing in `Python` *CSI Commun.* **23** 2

van Rossum G 1995 *Python Reference Manual* Technical Report. CWI (Centre for Mathematics and Computer Science), NLD

van Rossum G and Drake F L Jr 1995 *Python Tutorial* vol 620 (Amsterdam: Centrum voor Wiskunde en Informatica)

Vincent O R and Folorunso O 2009 A descriptive algorithm for Sobel image edge detection *Proc. of Informing Science and IT Education Conf. (InSITE)* vol 40 (Informing Science Institute) pp 97–107

White J M and Rohrer G D 1983 Image thresholding for optical character recognition and other applications requiring character image extraction *IBM J. Res. Dev.* **27** 400–11

Xu X, Xu S, Jin L and Song E 2011 Characteristic analysis of Otsu threshold and its applications *Pattern Recognit. Lett.* **32** 956–61

Chapter 8

Facial recognition

Facial recognition is one of the most prominent, challenging and long-attacked problems in artificial vision due to the multiple variables and character settings that are involved in this biometric. Facial recognition mainly consists of a four-phased problem. It includes face detection, face alignment, feature extraction and face classification. In this chapter we review '*classical*' face recognition algorithms then move on to review a more recent deep learning face recognition technique (Balaban 2015, Shanahan and Dai 2020).

8.1 Introduction to facial recognition

In terms of artificial vision, facial recognition has always been a challenging problem in computing. This is due to the various variants that solving the challenge may include, such as the viewing angles, illumination, facial expressions, facial hair, accessories (hats, glasses, etc) and age changing characteristics, to name a few. Some facial recognition algorithms identify facial features by extracting landmarks (or features) from an image of the subject's face. For example, an algorithm may analyse the relative position, size or shape of the eyes, nose, cheekbones and jaw. These features are then used to search for other images with matching or similar features.

The first recorded scientific attempt at successful facial recognition was performed by Goldstein in 1971 (Goldstein *et al* 1971) when he took into account 21 facial features that were manually measured to later identify a face in a photograph using a computer. This brute force algorithm proved to be an initial insight into how one could later extract facial features automatically.

In 1987 the *Eigenfaces algorithm* was presented by Sirovich and Kirby (Sirovich and Kirby 1978), which performs facial recognition when the faces of the dataset are aligned straight, looking forward and the pictures are cropped. This method proved to be encouraging by automatically extracting certain facial features using the images' eigenvectors which contained information about the eyes, mouth and nose. The dataset limitations were still apparent but the methods would continue to advance.

Adding to of the previous algorithms, machine learning algorithms (Zhang and Qiao 2003, Zhao *et al* 2015) started to appear in the early 2000s, with the *Haar cascade* being one of the most popular. Paul Viola and Michael Jones in their paper '*Rapid object detection using a boosted cascade of simple features*' (Viola and Jones 2001) popularized a quick method of obtaining facial recognition with a relatively high false positive rate. Other successful techniques include linear discriminant analysis (LDA) (Chen *et al* 2000) and local binary patterns (LBPs) (Brahnam *et al* 2013) for extracting key facial features.

In this chapter we will also apply a 'black-box' algorithm that uses a CNN for facial recognition (Hu *et al* 2015). CNNs have demonstrated continuous advances in object classification and detection accuracy (in their respective benchmark datasets) and inference speed (thanks to GPUs and different CNN architectures) (Khan *et al* 2020, Lawrence *et al* 1997, Li and Cha 2019, Liu *et al* 2016).

8.2 Local binary patterns for facial recognition

Relying on local feature extraction, we may probe a cropped image of a forward-looking face. In the 2004 paper by Ahonen, Hadid and Pietikainen that discusses facial recognition with LBPs (Ahonen *et al* 2004), they divide a frontal face image into an equally sized grid, obtain the histogram for each region in the grid, then use a weighting to give higher values to key facial features (cheeks, eyes, lips) then concatenate these histograms and use a nearest neighbour algorithm with a Chi-squared dissimilarity measure to tell them apart. They present three dissimilarity measures:

- Histogram intersection.
- Log-likelihood.
- The Chi-squared statistic.

As an example, we can divide a greyscale facial image using a 7×7 grid. We then compute the LBP histograms of each region and multiply them using an experimentally obtained (by tuning hyperparameters on the training data) weighted 7×7 grid canvas. The Chi-squared distance metric is then performed over the resulting matrix and a kNN algorithm with $k = 1$ will be used in the classification decision.

Usually LBPs tend to outperform the eigenface algorithms due to the higher abstraction of local facial features. Also, in contrast to the eigenfaces algorithm we do not have to compute the algorithm using the whole dataset again if a new face is to be added to the recognition task. Using `OpenCV`, we can implement this LBP algorithm. Specifically, we use the known CALTECH faces dataset.

First we import the necessary libraries and create a loading function (see listing 8.1 for the `Python` code). We then load the CALTECH faces dataset (see listing 8.2 for the `Python` code and figure 8.1 for the output).

8.3 The eigenfaces algorithm

Having as their main goal to distinguish faces that can be seen as well-defined patterns, Kirby and Sirovich used an empirical Karhunen–Loeve expansion, better

```python
from sklearn.utils import Bunch
from imutils import paths
from scipy import io
import numpy as np
import random
import cv2

def load_caltech_faces(datasetPath, min_faces=10, face_size=(47, 62), equal_samples=True,
        test_size=0.33, seed=42, flatten=False):
        # grab the image paths associated with the faces, then load the bounding box data
        imagePaths = sorted(list(paths.list_images(datasetPath)))
        bbData = io.loadmat("{}/ImageData.mat".format(datasetPath))
        bbData = bbData["SubDir_Data"].T
        random.seed(seed)
        data = []
        labels = []

        # loop over the image paths
        for (i, imagePath) in enumerate(imagePaths):
                # load the image and convert it to grayscale
                image = cv2.imread(imagePath)
                gray = cv2.cvtColor(image, cv2.COLOR_BGR2GRAY)

                # grab the bounding box associated with the current image, extract the face
                # ROI, and resize it to a canonical size
                k = int(imagePath[imagePath.rfind("_") + 1:][:4]) - 1
                (xBL, yBL, xTL, yTL, xTR, yTR, xBR, yBR) = bbData[k].astype("int")
                face = gray[yTL:yBR, xTL:xBR]
                face = cv2.resize(face, face_size)

# check if the face should be flattened into a single row
                if flatten:
                        face = face.flatten()

                # update the data matrix and associated labels
                data.append(face)
                labels.append(imagePath.split("/")[-2])

        # convert the data matrix and labels list to a NumPy array
```

Listing 8.1. `load_faces.py` implementation.

```
    data = np.array(data)
    labels = np.array(labels)

    # # check to see if equal samples for each face should be used
    if equal_samples:

        # initialize the list of sampled indexes
        sampledIdxs = []

        # loop over the unique labels
        for label in np.unique(labels):

            # grab the indexes into the labels array where labels equals the current
            # label

            labelIdxs = np.where(labels == label)[0]

            # only proceed if the required number of minimum faces can be met
            if len(labelIdxs) >= min_faces:



                labelIdxs = random.sample(list(labelIdxs), min_faces)
                sampledIdxs.extend(labelIdxs)

        # use the sampled indexes to select the appropriate data points and labels

        random.shuffle(sampledIdxs)
        data = data[sampledIdxs]
        labels = labels[sampledIdxs]

    # compute the training and testing split index

    idxs = range(0, len(data))
    random.shuffle(list(idxs))
    split = int(len(idxs) * (1.0 - test_size))

    # split the data into training and testing segments

    (trainData, testData) = (data[:split], data[split:])
```

Listing 8.1. (Continued.)

```
        (trainLabels, testLabels) = (labels[:split], labels[split:])

        # create the training and testing bunches

        training = Bunch(name="training", data=trainData, target=trainLabels)

        testing = Bunch(name="testing", data=testData, target=testLabels)

        # return a tuple of the training, testing bunches, and original labels

        return (training, testing, labels)
```

Listing 8.1. (Continued.)

```
from __future__ import print_function
from sklearn.preprocessing import LabelEncoder
from sklearn.metrics import classification_report

import numpy as np
import argparse
import imutils
import cv2

from load_faces import load_caltech_faces

ap = argparse.ArgumentParser()
ap.add_argument("-d", "--dataset", required=True, help="path␣to␣CALTECH␣Faces␣dataset")
ap.add_argument("-s", "--sample-size", type=int, default=10, help="#␣of␣example␣samples")
args = vars(ap.parse_args())

print("[INFO]␣loading␣CALTECH␣Faces␣dataset...")
(training, testing, names) = load_caltech_faces(args["dataset"], min_faces=21,
test_size=0.25)

le = LabelEncoder()
le.fit_transform(training.target)

recognizer = cv2.face.LBPHFaceRecognizer_create(radius=2, neighbors=16, grid_x=8, grid_y=8)

print("[INFO]␣training␣face␣recognizer...")
recognizer.train(training.data, le.transform(training.target))
```

Listing 8.2 `train_lbp.py` implementation.

```python
print("[INFO] gathering predictions...")
predictions = []
confidence = []

for i in range(0, len(testing.data)):

        # classify the face and update the list of predictions and confidence scores
        (prediction, conf) = recognizer.predict(testing.data[i])
        predictions.append(prediction)
        confidence.append(conf)

print(classification_report(le.transform(testing.target), predictions,
        target_names=np.unique(names)))
```

Listing 8.2 (Continued.)

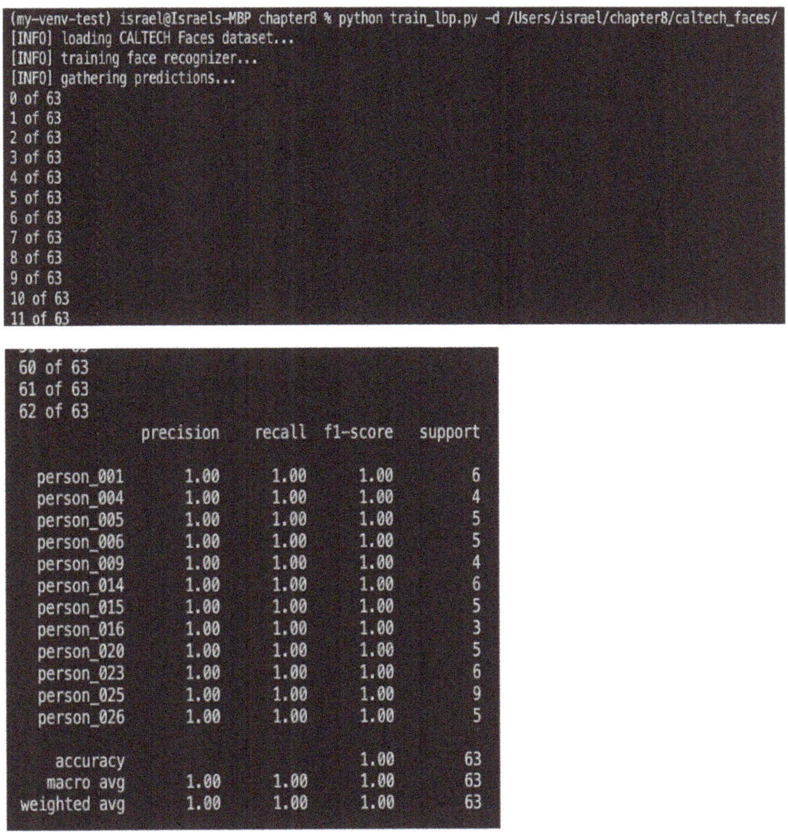

```
(my-venv-test) israel@Israels-MBP chapter8 % python train_lbp.py -d /Users/israel/chapter8/caltech_faces/
[INFO] loading CALTECH Faces dataset...
[INFO] training face recognizer...
[INFO] gathering predictions...
0 of 63
1 of 63
2 of 63
3 of 63
4 of 63
5 of 63
6 of 63
7 of 63
8 of 63
9 of 63
10 of 63
11 of 63
```

```
60 of 63
61 of 63
62 of 63
              precision    recall  f1-score   support

  person_001       1.00      1.00      1.00         6
  person_004       1.00      1.00      1.00         4
  person_005       1.00      1.00      1.00         5
  person_006       1.00      1.00      1.00         5
  person_009       1.00      1.00      1.00         4
  person_014       1.00      1.00      1.00         6
  person_015       1.00      1.00      1.00         5
  person_016       1.00      1.00      1.00         3
  person_020       1.00      1.00      1.00         5
  person_023       1.00      1.00      1.00         6
  person_025       1.00      1.00      1.00         9
  person_026       1.00      1.00      1.00         5

    accuracy                           1.00        63
   macro avg       1.00      1.00      1.00        63
weighted avg       1.00      1.00      1.00        63
```

Figure 8.1. Output for listing 8.2: terminal output of `train_lbp.py` using as the only argument the CALTECH faces dataset directory.

known as a principle component analysis (PCA). The Karhunen–Loeve expansion is analogue to a Fourier series expansion with the contrast that instead of using constants for the coefficients and sinusoidal functions for the expansion bases, the Karhunen–Loeve expansion uses random variables as the coefficients and the expansion bases are real-valued functions. In its analytic form the Karhunen–Loeve expansion may be seen in the following theorem (Sirovich and Kirby 1987):

Theorem 8.1. *Let $X = \{X_t\}$, $t \in [a, b] \subset \mathcal{R}$ be a centred mean-square continuous stochastic process with $X \in L^2$. There exists a basis $\{e_i\}$ of L^2 such that for all $t \in [a, b]$,*

$$X_t = \sum_{i=1}^{\infty} x_i e_i(t), \qquad \text{in } L^2, \tag{8.1}$$

where the coefficients x_i are given by $x_i(w) = \int_a^b X_t(w) e_i(t)\, dt$ and satisfy the following:

1. $E[x_i] = 0.$
2. $E[x_i x_j] = \delta_{ij}\lambda_j.$
3. $\mathrm{Var}[x_i] = \lambda_i.$

Now knowing that we can represent or decompose a face as a linear combination of *eigenfaces*, we can say that expanding up to a certain order (until a certain number of the expansion) can allow us to approximate a face using a set of *eigenfaces* with a relatively low error percentage. The practical implementation as seen in listing 8.4 concentrates on the expansion deviation from the mean as

$$\vec{\phi}^{(n)} = \vec{\psi}^{(n)} - \vec{\psi}_{\text{mean}}, \tag{8.2}$$

where $\phi^{(n)}$ is called a *caricature*, $\psi^{(n)}$ is a particular picture and $\bar{\psi}$ is the mean of the ensemble where $[\psi^{(n)}]$ denotes an ensemble of pictures for $n = 1, 2, 3, \ldots, N$. The practical implementation of this eigenfaces algorithm involves reducing the dimensionality where one can store the information about a face.

Consider the following correlation matrix:

$$\mathbf{C} = \frac{1}{M} \sum_{n=1}^{M} \phi^{(n)} \phi^{(m)}, \tag{8.3}$$

where $\mathbf{C}$ is symmetric and non-negative. Then its eigenvectors and eigenvalues are given by

$$\mathbf{C}\vec{u}^{(n)} = \lambda^{(n)} \vec{u}^{(n)}, \tag{8.4}$$

where $\vec{u}^{(n)}$ are orthonormal vectors such that the inner product is $(\vec{u}^{(n)}, \vec{u}^{(m)}) = \delta_{mn}$.

We then obtain the basis of expansion given by the eigenvectors $\vec{u}^{(n)}$. Thus we should be able to calculate any typical facial picture with a series of the form

$$\vec{\psi} = \vec{\psi}_{\text{mean}} + \sum_{n=1}^{M} a_n \vec{u}^{(n)}, \tag{8.5}$$

where $a_n = (\vec{u}^{(n)}, \vec{\psi} - \vec{\psi}_{\text{mean}})$.

The dimensionality is reduced if we compute fewer than M terms of the series. Likewise, one can still have a good approximation without having to use all of the terms of equation (8.5).

In short, the practical eigenfaces algorithm is characterized by the following steps:

1. Have an ensemble of training faces $\{\psi^{(n)}\}$, where $n = 1, 2, 3, \ldots, N$ and each facial image has dimension $M \times M$.
2. Concatenate each training face into a vector $\vec{\psi}^{(n)}$ of dimension $1 \times M^2$.
3. Compute the mean training face vector $\vec{\psi}_{\text{mean}}$ and obtain the deviation vectors $\vec{\phi}^{(n)} = \vec{\psi}^{(n)} - \vec{\psi}_{\text{mean}}$.
4. Calculate a covariance matrix $\mathbf{C} = \boldsymbol{\phi}\boldsymbol{\phi}^T$, where $\boldsymbol{\phi} = \left[\vec{\phi}^{(1)}\ \vec{\phi}^{(2)}\ \vec{\phi}^{(3)} \ldots \vec{\phi}^{(N)}\right]$.
5. Calculate the eigenvalues and eigenvectors of the covariance matrix $\mathbf{C}$. These eigenvectors may be called eigenfaces and will be the basis for the expansion of any new or existing faces.

Numerically, calculating the matrix $\mathbf{C}$ may be time consuming and non-optimal for real-time recognition. Thus usually when the size of the training images M is less than the dimension of $\mathbf{C}$ (in most cases) we can use another technique to obtain the eigenfaces. We start by computing the eigenvalue decomposition of $\boldsymbol{\phi}^T\boldsymbol{\phi}$ instead of $\boldsymbol{\phi}\boldsymbol{\phi}^T$,

$$\boldsymbol{\phi}^T\boldsymbol{\phi}\vec{u}^{(n)} = \lambda^{(n)}\vec{u}^{(n)} \tag{8.6}$$

and we can obtain

$$\boldsymbol{\phi}\boldsymbol{\phi}^T\boldsymbol{\phi}\vec{u}^{(n)} = \lambda^{(n)}\boldsymbol{\phi}\vec{u}^{(n)}, \tag{8.7}$$

which simplifies the extraction of eigenvectors by instead computing a lower dimensional matrix $\boldsymbol{\phi}^T\boldsymbol{\phi}$, where $\vec{v}^{(n)} = \boldsymbol{\phi}\vec{u}^{(n)}$ is an eigenvector of $\mathbf{C}$. We can use this technique due to the fact that $\mathbf{C}$ would be singular when $M < \dim \mathbf{C}$.

Factors that affect the error rate of this algorithm are the necessary use of square pictures and forward-looking faces, skin-tone, as well as strong effects due to possible fluctuations in lighting.

8.4 Example using the CALTECH faces dataset

We perform the algorithm on the CALTECH faces dataset, which is in ideal conditions. We first import the necessary libraries, then we load the CALTECH faces dataset and use PCA to visualize the eigenfaces with the `Python` script `resultsmontage.py` (see listing 8.3 for the code (Rosebrock 2017)).

```python
# import the necessary packages
import numpy as np
import cv2

class ResultsMontage:
        def __init__(self, imageSize, imagesPerRow, numResults):
                # store the target image size and the number of images per row
                self.imageW = imageSize[0]
                self.imageH = imageSize[1]
                self.imagesPerRow = imagesPerRow

                # allocate memory for the output image
                numCols = numResults // imagesPerRow
                self.montage = np.zeros((numCols * self.imageW, imagesPerRow * self.imageH, 3), dtype="uint8")

                # initialize the counter for the current image along with the row and column
                # number
                self.counter = 0
                self.row = 0
                self.col = 0

        def addResult(self, image, text=None, highlight=False):
                # check to see if the number of images per row has been met, and if so, reset
                # the column counter and increment the row
                if self.counter != 0 and self.counter % self.imagesPerRow == 0:
                        self.col = 0
                        self.row += 1

                # resize the image to the fixed width and height and set it in the montage
                image = cv2.resize(image, (self.imageH, self.imageW))
                (startY, endY) = (self.row * self.imageW, (self.row + 1) * self.imageW)
                (startX, endX) = (self.col * self.imageH, (self.col + 1) * self.imageH)
                self.montage[startY:endY, startX:endX] = image

                # if the text is not None, draw it
                if text is not None:
```

Listing 8.3 `resultsmontage.py` implementation.

```
                cv2.putText(self.montage, text, (startX + 10, startY + 30), cv2.FONT_HERSHEY_SIMPLEX,
                        1.0, (0, 255, 255), 3)

        # check to see if the result should be highlighted
        if highlight:
                cv2.rectangle(self.montage, (startX + 3, startY + 3), (endX - 3, endY - 3), (0, 255, 0), 4)

        # increment the column counter and image counter
        self.col += 1
        self.counter += 1
```

Listing 8.3 (Continued.)

Finally, to classify faces we may use an SVM (Hearst *et al* 1998, Guo *et al* 1970) or kNN algorithm using the Euclidean distance between projected eigenvector representations (see listing 8.4 for the code and figure 8.2 for the output).

We then observe the classification results. One of the biggest criticisms of the eigenfaces algorithm is the strict facial alignment required when training and identifying faces (see figure 8.3). Thanks to their robustness LBPs are usually in a better position for facial recognition but this always depends on practical external conditions.

Haar cascade

Although it is not state-of-the-art (Guo and Zhang 2019), we may use a Haar cascade algorithm (Sharifara *et al* 2014) which consists in extracting line, edge and four rectangle features from an image using convolutional kernels (Haar features).

To lower the amount of operations to calculate the Haar features a few steps need to be taken first. *Integral images* are introduced to reduce the number of operations performed, by making the convolutional operations over a cumulative image defined by the cumulative values of pixels at both the horizontal and vertical axes such that

$$I(x, y) = \sum_{x' \leqslant x, y' \leqslant y} i(x', y'), \tag{8.8}$$

where $I(x, y)$ is the integral image matrix and $i(x', y')$ is the image pixel intensity. The origin of the coordinates x', y' and x, y are at the upper left corner of each matrix. Then training is performed with a set of face and no-face images using all the Haar features. A threshold for the Haar features is calculated by minimizing classification error.

An iterative optimization algorithm called Adaboost is then used to modify the weights of each image until a minimum of classification is found. Each round of optimizations or each model found with different weights are called *weak learners*. We use the set of weak learners to make a strong learner as

```python
# import the necessary packages
from __future__ import print_function
from load_faces import load_caltech_faces
from resultsmontage import ResultsMontage
import sklearn
from sklearn.metrics import classification_report
from sklearn.svm import SVC
from skimage import exposure
import numpy as np
import argparse
import imutils
import cv2

# due to sklearn deprecation of RandomizedPCA this function will check the version
def is_sklearn_less_than_0_18():
        if int(sklearn.__version__.split(".")[1]) < 18:
                return True
        else:
                return False

# handle if sklearn is < 0.18 where we use RandomizedPCA
if is_sklearn_less_than_0_18():
        print("[INFO] sklearn=={}, so using RandomizedPCA".format(sklearn.__version__))
        from sklearn.decomposition import RandomizedPCA

# otherwise sklearn's RandomizedPCA is deprecated and we need to use PCA
else:
        print("[INFO] sklearn=={}, so using PCA".format(sklearn.__version__))
        from sklearn.decomposition import PCA

# construct the argument parse and parse command line arguments
ap = argparse.ArgumentParser()
ap.add_argument("-d", "--dataset", required=True, help="path to CALTECH Faces dataset")
ap.add_argument("-n", "--num-components", type=int, default=150, help="# of principal components")
ap.add_argument("-s", "--sample-size", type=int, default=5, help="# of example samples")
ap.add_argument("-v", "--visualize", type=int, default=-1,
        help="whether or not PCA components should be visualized")
args = vars(ap.parse_args())

# load the CALTECH faces dataset
```

Listing 8.4. eigenfaces.py implementation.

```
print("[INFO] loading CALTECH Faces dataset...")
(training, testing, names) = load_caltech_faces(args["dataset"], min_faces=21, flatten=True,
        test_size=0.25)

# compute the PCA (eigenfaces) representation of the data, then project the training data
# onto the eigenfaces subspace
print("[INFO] creating eigenfaces...")

# handle if sklearn is < 0.18
if is_sklearn_less_than_0_18():
        pca = RandomizedPCA(n_components=args["num_components"], whiten=True)

# otherwise sklearn is >= 0.18
else:
        pca = PCA(svd_solver="randomized", n_components=args["num_components"], whiten=True)

trainData = pca.fit_transform(training.data)

# check to see if the PCA components should be visualized
if args["visualize"] > 0:
        # initialize the montage for the components
        montage = ResultsMontage((62, 47), 4, 16)

        # loop over the first 16 individual components
        for (i, component) in enumerate(pca.components_[:16]):
                # reshape the component to a 2D matrix, then convert the data type to an unsigned
                # 8-bit integer so it can be displayed with OpenCV
                component = component.reshape((62, 47))
                component = exposure.rescale_intensity(component, out_range=(0, 255)).astype("uint8")
                component = np.dstack([component] * 3)
                montage.addResult(component)

        mean = pca.mean_.reshape((62, 47))
        mean = exposure.rescale_intensity(mean, out_range=(0, 255)).astype("uint8")
        cv2.imshow("Mean", mean)
        cv2.imshow("Components", montage.montage)
        cv2.waitKey(0)

# train a classifier on the eigenfaces representation
```

Listing 8.4. (Continued.)

```python
print("[INFO]␣training␣classifier...")
model = SVC(kernel="rbf", C=10.0, gamma=0.001, random_state=84)
model.fit(trainData, training.target)

# evaluate the model
print("[INFO]␣evaluating␣model...")
predictions = model.predict(pca.transform(testing.data))
print(classification_report(testing.target, predictions))

# loop over the desired number of samples
for i in np.random.randint(0, high=len(testing.data), size=(args["sample_size"],)):
        # grab the face and classify it
        face = testing.data[i].reshape((62, 47)).astype("uint8")
        prediction = model.predict(pca.transform(testing.data[i].reshape(1, -1)))

        # resize the face to make it more visable, then display the face and the prediction
        print("[INFO]␣Prediction:␣{},␣Actual:␣{}".format(prediction[0], testing.target[i]))
        face = imutils.resize(face, width=face.shape[1] * 2, inter=cv2.INTER_CUBIC)
        cv2.imshow("Face", face)
        cv2.waitKey(0)
```

Listing 8.4. (Continued.)

$$H(\mathbf{x}) = w_1h_1(\mathbf{x}) + w_2h_2(\mathbf{x}) + w_3h_3(\mathbf{x}) + \cdots + w_nh_n(\mathbf{x}), \tag{8.9}$$

where $H(\mathbf{x})$ is a strong learner, $h_1(\mathbf{x})$ is a weak learner, w_i for $i = 1, 2, 3, \ldots, n$ are weights and $\mathbf{x}$ is a feature vector.

Training samples that are misclassified are updated to have higher weight values. The full algorithm to obtain $H(\mathbf{x})$, if we use an input dataset

$$D = \{(\boldsymbol{x}_1, y_1), (\boldsymbol{x}_2, y_2), \cdots, (\boldsymbol{x}_m, y_m)\}$$

with a base learning algorithm $\mathcal{L}$ and a number of learning rounds T may be seen in the following algorithm:

1. Initialize the weight distribution
$$\mathcal{D}_1(i) = 1/m \quad .$$
2. Begin the iterative process for $t = 1, \cdots, T$.
3. Train a learner h_t from D using distribution $\mathcal{D}_t$ as
$$h_t = \mathcal{L}(D, \mathcal{D}_t).$$
4. Measure the error of h_t, with
$$\epsilon_t = Pr_{x\sim\mathcal{D}_t, y}\boldsymbol{I}[h_t(\boldsymbol{x}) \neq y].$$

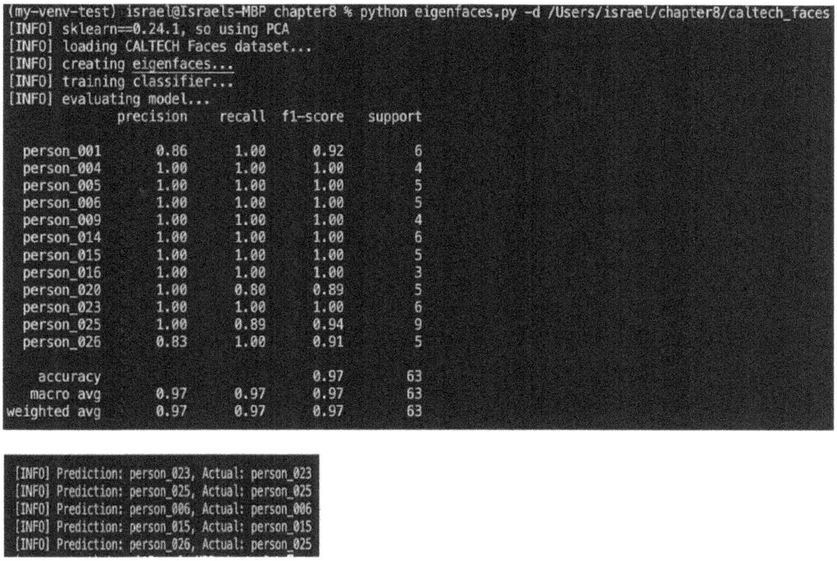

```
(my-venv-test) israel@Israels-MBP chapter8 % python eigenfaces.py -d /Users/israel/chapter8/caltech_faces
[INFO] sklearn==0.24.1, so using PCA
[INFO] loading CALTECH Faces dataset...
[INFO] creating eigenfaces...
[INFO] training classifier...
[INFO] evaluating model...
              precision    recall  f1-score   support

  person_001       0.86      1.00      0.92         6
  person_004       1.00      1.00      1.00         4
  person_005       1.00      1.00      1.00         5
  person_006       1.00      1.00      1.00         5
  person_009       1.00      1.00      1.00         4
  person_014       1.00      1.00      1.00         6
  person_015       1.00      1.00      1.00         5
  person_016       1.00      1.00      1.00         3
  person_020       1.00      0.80      0.89         5
  person_023       1.00      1.00      1.00         6
  person_025       1.00      0.89      0.94         9
  person_026       0.83      1.00      0.91         5

    accuracy                           0.97        63
   macro avg       0.97      0.97      0.97        63
weighted avg       0.97      0.97      0.97        63

[INFO] Prediction: person_023, Actual: person_023
[INFO] Prediction: person_025, Actual: person_025
[INFO] Prediction: person_006, Actual: person_006
[INFO] Prediction: person_015, Actual: person_015
[INFO] Prediction: person_026, Actual: person_025
```

Figure 8.2. Output of listing 8.4: terminal output of `eigenfaces.py` using as the only argument the CALTECH faces dataset directory.

Figure 8.3. The different faces of the people that were to be recognized in the `eigenfaces.py` output.

5. If $\epsilon_t > 0.5$ then break.
6. Determine the weight of h_t, with

$$\alpha_t = \frac{1}{2} \ln\left(\frac{1 - \epsilon_t}{\epsilon_t}\right).$$

7. Update the distribution, where Z_t is a normalization constant which enables $\mathcal{D}_{t+1}$ to be a distribution

$$\begin{aligned} \mathcal{D}_{t+1}(i) &= \frac{\mathcal{D}_t(i)}{Z_t} \times \begin{cases} \exp(-\alpha_t) & \text{if } h_t(\boldsymbol{x}_i) = y_i \\ \exp(\alpha_t) & \text{if } h_t(\boldsymbol{x}_i) \neq y_i \end{cases} \\ &= \frac{\mathcal{D}_t(i)\exp\left(-\alpha_t y_i h_t(\boldsymbol{x}_i)\right)}{Z_t}. \end{aligned}$$

8. Output

$$H(\boldsymbol{x}) = \text{sign}\left(\sum_{t=1}^{T} \alpha_t h_t(\boldsymbol{x})\right).$$

A lot of operations still need to be calculated before facial features are obtained for an image. Thus to speed the process up, a *cascade of classifiers* that act like a facial detector are used. The cascade of classifiers implies we calculate features by stages and each stage has a certain growing number of features to be calculated. Thus if a certain region of an image does not pass the first stage of feature verification it is said to not contain a face and so the rest of the features are not calculated.

8.4.1 Create a personal dataset

Using a Haar cascade algorithm embedded into an `OpenCV` video programme we can extract various cropped photos of our face (in a constricted frontal manner) in order to have training pictures of ourselves. The code is presented in listing 8.5.

The `-face-cascade` file is a pretrained Haar cascade xml file that is included in the `OpenCV` library. One may download it from the `OpenCV` Github. Then we run the programme using the shell commands in listing 8.6.

We then serialize these frames in order to reduce the memory requirements. We can deserialize a frame with the code in listing 8.6.1.

```
# import the necessary packages
from __future__ import print_function
from face_recognition import FaceDetector
from imutils import encodings
import argparse
import imutils
import cv2

# construct the argument parse and parse command line arguments
ap = argparse.ArgumentParser()
ap.add_argument("-f", "--face-cascade", required=True, help="path to face detection cascade")
ap.add_argument("-o", "--output", required=True, help="path to output file")
ap.add_argument("-w", "--write-mode", type=str, default="a+", help="write method for the output file")
args = vars(ap.parse_args())

# initialize the face detector, boolean indicating if we are in capturing mode or not, and
# the bounding box color
fd = FaceDetector(args["face_cascade"])
captureMode = False
```

Listing 8.5 Implementation of `gather_selfies.py`.

```
color = (0, 255, 0)

# grab a reference to the webcam and open the output file for writing
camera = cv2.VideoCapture(0)
f = open(args["output"], args["write_mode"])
total = 0
# loop over the frames of the video
while True:
        # grab the current frame
        (grabbed, frame) = camera.read()

        # if the frame could not be grabbed, then we have reached the end of the video
        if not grabbed:
                break

        # resize the frame, convert the frame to grayscale, and detect faces in the frame
        frame = imutils.resize(frame, width=500)
        gray = cv2.cvtColor(frame, cv2.COLOR_BGR2GRAY)
        faceRects = fd.detect(gray, scaleFactor=1.1, minNeighbors=9, minSize=(100, 100))

        # ensure that at least one face was detected
        if len(faceRects) > 0:
                # sort the bounding boxes, keeping only the largest one
                (x, y, w, h) = max(faceRects, key=lambda b:(b[2] * b[3]))

                # if we are in capture mode, extract the face ROI, encode it, and write it to file
                if captureMode:
                        face = gray[y:y + h, x:x + w].copy(order="C")
                        f.write("{}\n".format(encodings.base64_encode_image(face)))
                        total += 1

                # draw bounding box on the frame
                cv2.rectangle(frame, (x, y), (x + w, y + h), color, 2)

        # show the frame and record if the user presses a key
        cv2.imshow("Frame", frame)
        key = cv2.waitKey(1) & 0xFF

```

Listing 8.5 (Continued.)

```
    # if the 'c' key is pressed, then go into capture mode
    if key == ord("c"):
        # if we are not already in capture mode, drop into capture mode
        if not captureMode:
            captureMode = True
            color = (0, 0, 255)
        else:
            captureMode = False
            color = (0, 255, 0)

    elif key == ord("q"):
        break

print("[INFO]␣wrote␣{}␣frames␣to␣file".format(total))
f.close()
camera.release()
cv2.destroyAllWindows()
```

Listing 8.5 (Continued.)

```
$python gather_selfies.py --face-cascade cascades/haarcascade_frontalface_default.xml \
--output output/faces/israel.txt
```

Listing 8.5.1 Execution of `gather_selfies.py`.

```
$cd output/faces
$python
>>> from imutils import encodings
>>> import cv2
>>> f = open("israel.txt")
>>> l = f.readline().strip()
>>> frame = encodings.base64_decode_image(l)
>>> cv2.imshow("Frame", frame)
```

Listing 8.5.2 Code run after execution of `gather_selfies.py` for verification.

```
# import the necessary packages
from collections import namedtuple
import cv2
import imutils
import os
import pickle
```

Listing 8.6 `facerecognizer.py`.

```python
# define the face recognizer instance
FaceRecognizerInstance = namedtuple("FaceRecognizerInstance", ["trained", "labels"])

class FaceRecognizer:
    def __init__(self, recognizer, trained=False, labels=None):
        # store the face recognizer, whether or not the face recognizer has already been trained,
        # and the list of face name labels
        self.recognizer = recognizer
        self.trained = trained
        self.labels = labels

    def setLabels(self, labels):
        # store the face name labels
        self.labels = labels

    def setConfidenceThreshold(self, confidenceThreshold):
        # set the confidence threshold for the classifier
        if imutils.is_cv2():
            self.recognizer.setDouble("threshold", confidenceThreshold)
        else:
            self.recognizer.setThreshold(confidenceThreshold)

    def train(self, data, labels):
        # if the model has not been trained, train it
        if not self.trained:
            self.recognizer.train(data, labels)
            self.trained = True
            return

        # otherwise, update the model
        self.recognizer.update(data, labels)

    def predict(self, face):
        # predict the face
        (prediction, confidence) = self.recognizer.predict(face)

        # if the prediction is `-1`, then the confidence is greater than the threshold, implying
        # that the face cannot be recognized
        if prediction == -1:
            return ("Unknown", 0)

```

Listing 8.6.1 Implementation of `facerecognizer.py`.

```
        # return a tuple of the face label and the confidence
        return (self.labels[prediction], confidence)

    def save(self, basePath):
        # construct the face recognizer instance
        fri = FaceRecognizerInstance(trained=self.trained, labels=self.labels)

        # due to strange behavior with OpenCV, we need to make sure the output classifier file
        # exists prior to writing it to file
        if not os.path.exists(basePath + "/classifier.model"):
            f = open(basePath + "/classifier.model", "w")
            f.close()


        self.recognizer.write(basePath + "/classifier.model")

        # write the parameters to file
        f = open(basePath + "/fr.cpickle", "wb")
        f.write(pickle.dumps(fri))
        f.close()

    @staticmethod
    def load(basePath):
        # load the face recognition instance and construct the OpenCV face recognizer
        fri = pickle.loads(open(basePath + "/fr.cpickle", "rb").read())

        # handle if we are building an OpenCV 2.4 face recognizer
        if imutils.is_cv2():
            recognizer = cv2.createLBPHFaceRecognizer()
            recognizer.load(basePath + "/classifier.model")
        # otherwise we are building an OpenCV 3 face recognizer
        else:
            recognizer = cv2.face.LBPHFaceRecognizer_create()
            recognizer.read(basePath + "/classifier.model")
        # construct and return the face recognizer
        return FaceRecognizer(recognizer, trained=fri.trained, labels=fri.labels)
```

Listing 8.6.2 (Continued.)

8.5 A LBP face recognizer for your own face

The reason to use LBPs for face recognition is the accuracy and speed robustness with respect to a Haar cascade (Adouani *et al* 2019). We can first start off by defining a recognizer class that sets the labels and trains our recognizer. Here there are two arguments:

- *Data*: The list of image ROIs containing the face. The data list will be constructed from our set of serialized training images.
- *Labels*: The labels are simply the names of the people associated with the faces in the data list.

We then construct a predict function to begin the facial recognition process.

Finally, let us define a save and load method, so we can serialize our FaceRecognizer to the disk and restore it at a later time, using the code in listing 8.6.

We now show the algorithm for face detector functionality in listing 8.7.

Then we can use `train_recognizer.py` to train our model. We include `train_recognizer.py` in the code in listing 8.8.

Then we use the face recognizer `recognize.py` on a connected camera or webcam opened with `OpenCV`'s VideoCapture camera 0 (see listing 8.9 for the code).

```
# import the necessary packages
import cv2
import imutils

class FaceDetector:
        def __init__(self, faceCascadePath):
                # load the face detector
                self.faceCascade = cv2.CascadeClassifier(faceCascadePath)

        def detect(self, image, scaleFactor=1.1, minNeighbors=5, minSize=(30, 30)):
                # detect faces in the image
                flags = cv2.cv.CV_HAAR_SCALE_IMAGE if imutils.is_cv2() else cv2.CASCADE_SCALE_IMAGE
                rects = self.faceCascade.detectMultiScale(image, scaleFactor=scaleFactor,
                        minNeighbors=minNeighbors, minSize=minSize, flags=flags)

                # return the bounding boxes around the faces in the image
                return rects
```

Listing 8.7 Implementation of `facedetector.py`.

```python
# USAGE
# python train_recognizer.py --selfies output/faces --classifier output/classifier --sample-size 100

# import the necessary packages
from __future__ import print_function
from face_recognition.facerecognizer import FaceRecognizer
from imutils import encodings
import numpy as np
import argparse
import imutils
import random
import glob
import cv2

# construct the argument parse and parse command line arguments
ap = argparse.ArgumentParser()
ap.add_argument("-s", "--selfies", required=True, help="path␣to␣the␣selfies␣directory")
ap.add_argument("-c", "--classifier", required=True, help="path␣to␣the␣output␣classifier␣directory")
ap.add_argument("-n", "--sample-size", type=int, default=100, help="maximum␣sample␣size␣for␣each␣face")
args = vars(ap.parse_args())

# initialize the face recognizer and the list of labels

fr = FaceRecognizer(cv2.face.LBPHFaceRecognizer_create(radius=1, neighbors=8,
        grid_x=8, grid_y=8))

# initialize the list of labels
labels = []

# loop over the input faces for training
for (i, path) in enumerate(glob.glob(args["selfies"] + "/*.txt")):
        # extract the person from the file name,
        name = path[path.rfind("/") + 1:].replace(".txt", "")
        print("[INFO]␣training␣on␣'{}'".format(name))

        # load the faces file, sample it, and initialize the list of faces
        sample = open(path).read().strip().split("\n")
        sample = random.sample(sample, min(len(sample), args["sample_size"]))
        faces = []

```

Listing 8.8 Implementation of `train_recognizer.py`.

```
        # loop over the faces in the sample
        for face in sample:
                # decode the face and update the list of faces
                faces.append(encodings.base64_decode_image(face))

        # train the face detector on the faces and update the list of labels
        fr.train(faces, np.array([i] * len(faces)))
        labels.append(name)

# update the face recognizer to include the face name labels, then write the model to file
fr.setLabels(labels)
fr.save(args["classifier"])
```

Listing 8.8 (Continued.)

```
# import the necessary packages
from face_recognition.facedetector import FaceDetector
from face_recognition.facerecognizer import FaceRecognizer
import argparse
import imutils
import cv2

# construct the argument parse and parse command line arguments
ap = argparse.ArgumentParser()
ap.add_argument("-f", "--face-cascade", required=True, help="path to face detection cascade")
ap.add_argument("-c", "--classifier", required=True, help="path to the classifier")
ap.add_argument("-t", "--confidence", type=float, default=100.0,
        help="maximum confidence threshold for positive face identification")
args = vars(ap.parse_args())

# initialize the face detector, load the face recognizer, and set the confidence
# threshold
fd = FaceDetector(args["face_cascade"])
fr = FaceRecognizer.load(args["classifier"])
fr.setConfidenceThreshold(args["confidence"])
```

Listing 8.9 Implementation of `recognizer.py`.

```
# grab a reference to the webcam
camera = cv2.VideoCapture(0)

# loop over the frames of the video
while True:
    # grab the current frame
    (grabbed, frame) = camera.read()

    # if the frame could not be grabbed, then we have reached the end of the video
    if not grabbed:
        break

    # resize the frame, convert the frame to grayscale, and detect faces in the frame
    frame = imutils.resize(frame, width=500)
    gray = cv2.cvtColor(frame, cv2.COLOR_BGR2GRAY)
    faceRects = fd.detect(gray, scaleFactor=1.1, minNeighbors=5, minSize=(100, 100))
    # loop over the face bounding boxes
    for (i, (x, y, w, h)) in enumerate(faceRects):
        # grab the face to predict
        face = gray[y:y + h, x:x + w]

        # predict who's face it is, display the text on the image, and draw a bounding
        # box around the face
        (prediction, confidence) = fr.predict(face)
        prediction = "{}: {:.2f}".format(prediction, confidence)
        cv2.putText(frame, prediction, (x, y - 20), cv2.FONT_HERSHEY_SIMPLEX, 0.75, (0, 255, 0), 2)
        cv2.rectangle(frame, (x, y), (x + w, y + h), (0, 255, 0), 2)

    # show the frame and record if the user presses a key
    cv2.imshow("Frame", frame)
    key = cv2.waitKey(1) & 0xFF

    # if the 'q' key is pressed, break from the loop
    if key == ord("q"):
        break

# cleanup the camera and close any open windows
camera.release()
cv2.destroyAllWindows()
```

Listing 8.9 (Continued.)

8.6 Deep learning facial recognition

The general pipeline for facial recognition is bi-phased. First, localize and extract the face and then classify it accordingly to the person whose face it belongs to. There exist many algorithms where facial features are manually extracted but CNNs and other deep learning methods have obtained state-of-the-art precision and accuracy if using a large enough dataset for supervised training and a viable enough validation set (Sandler *et al* 2018).

The approach necessary to use CNNs includes submitting an input training dataset containing thousands of image examples of each class to then statistically generalize every class up to a certain level of abstraction. The main concept with these algorithms is based on classical inferential statistics: the larger the size of a training dataset is, the stronger the neural network's ability to be able to abstract common features and predict a never-before seen image. Each layer in these neural networks may learn certain inherent characteristics such as colour, texture and positional invariance as well as other local features. The deeper (more layers) a neural network, the more abstract features it can pick up. Making the neural network too deep may result in a loss of generalization and overfitting the model (Nielsen 2018). Having a shallow neural network will also result in not extracting enough features or the more complex features that are necessary to solve the imposed problem. This is why different architectures are created for different classification or detection problems. Specifically, the combination of these higher level abstractions allows solving facial recognition with unprecedented stability.

In 2014 DeepFace (Taigman *et al* 2014) achieved state-of-the-art accuracy on the famous Labelled Faces in the Wild (LFW) benchmark dataset, approaching human performance on the unconstrained condition for the first time (DeepFace: 97.35% versus human: 97.53%), by training a nine-layer model on 4 million facial images.

The mainstream network architectures, such as DeepFace, DeepID series (Sun *et al* 2015), VGGFace, FaceNet (Schroff *et al* 2015), VGGFace2 and other architectures (Ranjan *et al* 2019) are designed specifically for obtaining facial features and for use in facial recognition. The majority of the findings of these models are new features in the architectures or calculations in each perceptron. Overall, facial recognition accuracy is more heavily dependent on the training, testing and validation dataset consistency and quantity more than on the network architecture. Most of the recent facial recognition architectures have relatively small changes to their algorithms and so usually do not represent such a large relative change in accuracy for model performance (Wang and Deng 2021).

One can see comparisons and analysis on public databases that are of vital importance for both model training and testing. Major facial benchmarks are contained in the datasets such as LFW, IARPA Janus Benchmark-A/B/C face challenges (IJB-A/B/C), Megaface and MSCeleb-1M (Trigueros *et al* 2018). These datasets can be reviewed and compared in terms of the four aspects, training methodology, evaluation tasks and metrics, and recognition scenes, which provide useful references for training and testing deep face recognition.

Three modules are typically needed for a facial recognition automatic system. First, a face detector is applied to localize faces in images or videos. For a robust system the face detector should be capable of detecting faces with varying pose, illumination and scale. Also, the locations and sizes of the face bounding boxes should be precisely determined so that they contain a minimal amount of background. Second, a detector is used to localize the important facial landmarks, such as the eye centres, nose tip, mouth corners, etc. These points are used to align the faces to normalized coordinates to minimize the effects of rotation and scaling.

Third, a feature descriptor that encodes the identity information is extracted from the aligned face. Given the face representations, similarity scores are then obtained using a metric. If this metric is lower than a threshold, it signifies that the two faces are from the same subject. Since the early 1990s many face identification/verification algorithms have been shown to work well on images and videos that are collected in controlled settings (Tan *et al* 2006, Guo *et al* 1970, Chen *et al* 2000).

However, the performance of these algorithms often degrades significantly on face images or videos that have large variations in pose, illumination, resolution, expression, age, background clutter and occlusion. Moreover, for the application of video surveillance where a subject needs to be identified from hundreds of low resolution videos, the algorithm should be fast and robust.

8.6.1 Face extraction

For face extraction (obtaining the bounding box coordinates that indicate where the face is) various algorithms exist which we may use depending on our situation. If we only have people looking straight forward in a picture a histogram of gradients (HOG) (Dalal and Triggs 2005, Shu *et al* 2011) for feature extraction and an SVM for training different faces may be directly implemented (Li and Jain 2011, Zhang and Qiao 2003). For faster performance but overall higher false positive rate, the Haar cascade is the algorithm we would go with (Wilson and Fernandez 2006, Adouani *et al* 2019). Finally, the method that we will see because of its overall performance for several face profiles is a deep neural network (DNN) method (Ranjan *et al* 2018).

A practical method for facial recognition is to implement a pretrained DNN which is already contained in the `OpenCV` library (Bradski and Kaehler 2013). We can make use of it to extract faces instead of the Haar cascade, as in the previous subsection of this chapter.

This DNN module is a single-shot detector (SSD) (Liu *et al* 2016) built using a ResNet architecture trained on the Caffe framework. The main contributor Aleksandr Rybnikov has left this module in the `OpenCV` Github repository: opencv/3rdparty/dnn/samples/face_detector/ and in opencv/samples/dnn/face_detector/. In this case we need a `.prototxt` file which defines the network architecture and a `.caffemodel` file which contains the pretrained weights.

The `.prototext` file may be found at https://raw.githubusercontent.com/opencv/opencv/master/samples/dnn/face_detector/deploy.prototxt and the `.caffemodel` file may be found at raw.githubusercontent.com/opencv/opencv_3rdparty/dnn_samples_-face_detector_20180205_fp16/res10_300x300_ssd_iter_140000_fp16.caffemodel.

```python
import numpy as np
import argparse
import cv2

# we parse the necessary arguments

ap = argparse.ArgumentParser()
ap.add_argument("-m", "--model", help="path to Caffe model file", default = "res10_300x300_ssd_iter_140000_fp16.caffemodel")
ap.add_argument("-p", "--prototxt", help="path to Caffe prototxt file", default = "deploy.prototxt")
ap.add_argument("-i", "--image", required=True, help="path of the image to do facial recognition on")
```


```python
ap.add_argument("-i", "--image", required=True, help="path of the image to do facial recognition on")
```


```python
ap.add_argument("-c", "--confidence", type=float, default=0.2, help="confidence C for recognition threshold",)

args = vars(ap.parse_args())

print("[INFO] loading model...")
net = cv2.dnn.readNetFromCaffe(args["prototxt"], args["model"])
# load the input image and construct an input blob for the image
# by resizing to a fixed 300x300 pixels and then normalizing it
image = cv2.imread(args["image"])
(h, w) = image.shape[:2]
blob = cv2.dnn.blobFromImage(cv2.resize(image, (300, 300)), 1.0,
        (300, 300), (104.0, 177.0, 123.0))


print("[INFO] computing object detections...")
net.setInput(blob)
detections = net.forward()

# loop over the detections
for i in range(0, detections.shape[2]):
        # extract the confidence (i.e., probability) associated with the
        # prediction
        confidence = detections[0, 0, i, 2]
        # filter out weak detections by ensuring the ‘confidence’ is
        # greater than the minimum confidence
        if confidence > args["confidence"]:
```

Listing 8.10 Implementation of dnn_face_recognition.py.

```
            # compute the (x, y)-coordinates of the bounding box for the
            # object
            box = detections[0, 0, i, 3:7] * np.array([w, h, w, h])
            (startX, startY, endX, endY) = box.astype("int")

            # draw the bounding box of the face along with the associated
            # probability
            text = "{:.2f}%".format(confidence * 100)
            y = startY - 10 if startY - 10 > 10 else startY + 10
            cv2.rectangle(image, (startX, startY), (endX, endY),
                (0, 0, 255), 2)
            cv2.putText(image, text, (startX, y),
                cv2.FONT_HERSHEY_SIMPLEX, 0.45, (0, 0, 255), 2)
# show the output image
cv2.imshow("Output", image)
cv2.waitKey(0)
```

Listing 8.10 (Continued.)

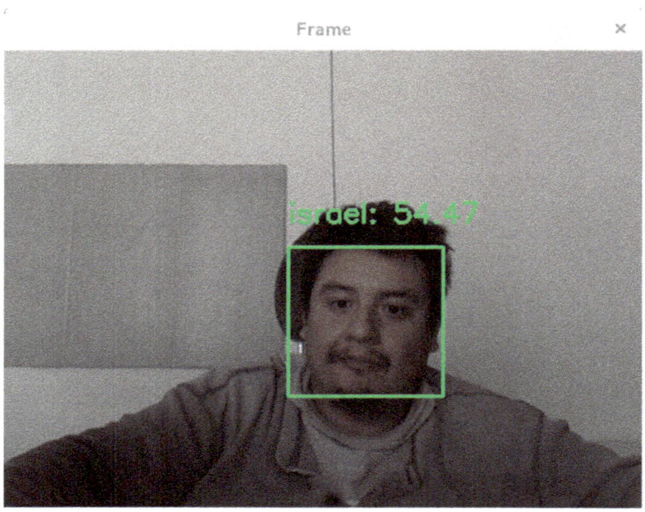

Figure 8.4. Terminal output of `FaceRecognition.py`.

After downloading these two files, we then start by importing the libraries and loading the neural network model. Then the class `cv2.blobFromImage` basically formats any input image by normalizing and resizing for model inference. We then pass the image forward through the model and we obtain the coordinates for our bounding box with a minimum threshold C (see listing 8.10 for the code and figure 8.4 for the output).

8.7 End notes

In this chapter we reviewed the most important facial recognition algorithms of the twenty-first century. The bases for facial feature extraction were once obtained in these semi-convolutional and statistical methods such as the Haar cascade, which could eventually transform into full CNN models and pretrained weights for popular use. The advances in lightweight architectures, SSDs and residual networks have permitted state-of-the-art facial detection accuracy and reduced facial detection times. This together with the new metric algorithms that have helped in facial recognition and differentiation between different images, have made deep learning CNN methods the best performing and flexible to date.

References

Adouani A, Ben Henia W M and Lachiri Z 2019 Comparison of Haar-like, HOG and LBP approaches for face detection in video sequences *16th Int. Multi-Conf. on Systems, Signals Devices (SSD)* pp 266–71

Ahonen T, Hadid A and Pietikäinen M 2004 Face recognition with local binary patterns *Computer Vision - ECCV 2004* ed T Pajdla and J Matas (Lecture Notes in Computer Science, vol 3021) (Berlin: Springer)

Balaban S 2015 Deep learning and face recognition: the state of the art *Proc. SPIE* **9457** 94570B

Bradski G 2013 *Learning OpenCV: Computer Vision in C++ with the OpenCV Library* 2nd edn (Sebasopol, CA: O'Reilly Media, Inc.)

Brahnam S, Jain L C, Nanni L and Lumini A 2013 *Local Binary Patterns: New Variants and Applications* (Berlin: Springer)

Chen L-F, Liao H-Y M, Ko M-T, Lin J-C and Yu G-J 2000 A new LDA-based face recognition system which can solve the small sample size problem *Pattern Recognit.* **33** 1713–26

Dalal N and Triggs B 2005 Histograms of oriented gradients for human detection *IEEE Computer Society Conf. on Computer Vision and Pattern Recognition* vol 1 pp 886–93

Goldstein A J, Harmon L D and Lesk A B 1971 Identification of human faces *Proc. IEEE* **59** 748–60

Guo G and Zhang N 2019 A survey on deep learning based face recognition *Comput. Vis. Image Underst.* **189** 102805

Guo G, Li S and Chan K 1970 Face recognition by support vector machines *Proc. of the Fourth IEEE Int. Conf. on Automatic Face and Gesture Recognition* pp 196–201

Hearst M A, Dumais S T, Osuna E, Platt J and Scholkopf B 1998 Support vector machines *IEEE Intell. Syst. Appl.* **13** 18–28

Hu G, Yang Y, Yi D, Kittler J, Christmas W, Li S Z and Hospedales T 2015 *When Face Recognition Meets with Deep Learning: An Evaluation of Convolutional Neural Networks for Face Recognition* **arXiv:**1504.02351

Khan A, Sohail A, Zahoora U and Qureshi A S 2020 A survey of the recent architectures of deep convolutional neural networks *Artif. Intell. Rev.* **53** 5455–516

Lawrence S, Giles C L, Tsoi A C and Back A D 1997 Face recognition: a convolutional neural-network approach *IEEE Trans. Neural Netw.* **8** 98–113

Li S Z and Jain A K 2011 *Handbook of Face Recognition* 2nd edn (Berlin: Springer)

Li Y and Cha S 2019 *Face Recognition System* arXiv:1901.02452

Liu W, Anguelov D, Erhan D, Szegedy C, Reed S, Fu C-Y and Berg A C 2016 *SSD: Single Shot MultiBox Detector* (Lecture Notes in Computer Science) (Berlin: Springer) pp 21–37

Nielsen M A 2018 *Neural Networks and Deep Learning* (http://neuralnetworksanddeeplearning.com/)

Ranjan R, Patel V M and Chellappa R 2019 HyperFace: a deep multi-task learning framework for face detection, landmark localization, pose estimation, and gender recognition *IEEE Trans. Pattern Anal. Mach. Intell.* **41** 121–35

Ranjan R, Sankaranarayanan S, Bansal A, Bodla N, Chen J C, Patel V M, Castillo C D and Chellappa R 2018 Deep learning for understanding faces: machines may be just as good, or better, than humans *IEEE Signal Process. Mag.* **35** 66–83

Rosebrock M A 2017 *Deep Learning for Computer Vision with* `Python`*: Starter Bundle* (PyImageSearch)

Sandler M, Howard A, Zhu M, Zhmoginov A and Chen L-C 2018 Mobilenetv2: inverted residuals and linear bottlenecks *Proc. IEEE Conf. on Computer Vision and Pattern Recognition* pp 4510–20

Schroff F, Kalenichenko D and Philbin J 2015 FaceNet: a unified embedding for face recognition and clustering *IEEE Conf. on Computer Vision and Pattern Recognition* (Washington, DC: IEEE Computer Society) pp 815–23

Shanahan J G and Dai L 2020 Introduction to computer vision and real time deep learning-based object detection *Proc. of the 26th ACM SIGKDD Int. Conf. on Knowledge Discovery and Data Mining (New York)* (New York: Association for Computing Machinery) pp 3523–4

Sharifara A, Mohd Rahim M S and Anisi Y 2014 A general review of human face detection including a study of neural networks and Haar feature-based cascade classifier in face detection *2014 Int. Symp. on Biometrics and Security Technologies (ISBAST)* pp 73–8

Shu C, Ding X and Fang C 2011 Histogram of the oriented gradient for face recognition *Tsinghua Sci. Technol.* **16** 216–24

Sirovich L and Kirby M 1987 Low-dimensional procedure for the characterization of human faces *J. Opt. Soc. Am. A* **4** 519–24

Sun Y, Liang D, Wang X and Tang X 2015 DeepID3: face recognition with very deep neural networks arXiv:1502.00873

Taigman Y, Yang M and Ranzato M' A 2014 DeepFace: closing the gap to human-level performance in face verification *Proc. of the IEEE Conf. on Computer Vision and Pattern Recognition (CVPR)* pp 1701–8

Tan X, Chen S, Zhou Z-H and Zhang F 2006 Face recognition from a single image per person: a survey *Pattern Recognit.* **39** 1725–45

Trigueros D S, Meng L and Hartnett M 2018 Face recognition: from traditional to deep learning methods arXiv:1811.00116

Viola P A and Jones M J 2001 Rapid object detection using a boosted cascade of simple features *Proc. of the 2001 IEEE Computer Society Conf. on Computer Vision and Pattern Recognition (CVPR 2001)* pp 511–8

Wang M and Deng W 2021 Deep face recognition: a survey *Neurocomputing* **429** 215–44

Wilson P I and Fernandez J 2006 Facial feature detection using Haar classifiers *J. Comput. Sci. Coll.* **21** 127–33

Zhang S and Qiao H 2003 Face recognition with support vector machine *IEEE Int. Conf. on Robotics, Intelligent Systems and Signal Processing* vol 2 pp 726–30

Zhao C, Li X and Cang Y 2015 Bisecting *k*-means clustering based face recognition using block-based bag of words model *Optik* **126** 1761–6

Rafael G González-Acuña, Héctor A Chaparro-Romo and Israel Melendez-Montoya

Chapter 9

Artificial vision case studies

In the previous chapters we have studied the theoretical backgrounds of popular classical and deep learning artificial vision algorithms. We also implemented popular deep learning algorithms for application tasks. The code examples that were shown in the previous chapters were used as learning examples for the common implementation of the artificial vision algorithms. In this chapter we will use state-of-the-art implementations for specific artificial vision projects that may be useful for the reader. We hope these project applications may serve as algorithm baselines for custom production `models`.

9.1 Measuring the camera–object distance

There exist multiple methods to measure the distance from a camera's CCD or CMOS sensor to an object. The most simple concept using a classical method consists in having a reference width object and distance in order to calibrate a camera in terms of *true* focal distance. We will review two methods for estimating the camera–object distance.

9.1.1 Camera distortion calibration

Economy cameras can be used in order to roughly estimate the distance from a camera's sensor to an object. The precision obtained with this method may not be great but will serve as a quick reference. The 2D distortion induced by economy cameras may be corrected up to a certain point using `OpenCV`'s `cv2.findChessboardCorners`, as we can see in this subsection.

The radial and tangential distortion induced by a pinhole camera may be modelled using the Brown–Conrady model (Brown 1966, Brown 1971, Pozdnyakov 2020) with the following equations:

$$\begin{aligned} x_{\mathrm{u}} = \; & x_{\mathrm{d}} + (x_{\mathrm{d}} - x_{\mathrm{c}})(K_1 r^2 + K_2 r^4 + K_3 r^6 + \cdots) + P_1(r^2 + 2(x_{\mathrm{d}} - x_{\mathrm{c}})^2) \\ & + 2P_2(x_{\mathrm{d}} - x_{\mathrm{c}})(y_{\mathrm{d}} - y_{\mathrm{c}})(1 + P_3 r^2 + P_4 r^4 + P_5 r^6 + \cdots), \end{aligned} \tag{9.1}$$

doi:10.1088/978-0-7503-3707-6ch9

$$y_u = y_d + (y_d - y_c)(K_1r^2 + K_2r^4 + K_3r^6 + \cdots) + P_1(x_d - x_c)(y_d - y_c) + 2P_2(r^2 + 2(y_d - y_c)^2)(1 + P_3r^2 + P_4r^4 + P_5r^6 + \cdots), \tag{9.2}$$

where x_u, y_u is an undistorted point, x_d, y_d is the distorted point, x_c, y_c is the distortion centre point, K_n and P_n are the radial distortion coefficient and tangential distortion coefficient, respectively, using $n = 1, 2, 3, \ldots,$ and $r = \sqrt{(x_d - x_c)^2 + (y_d - y_c)^2}$.

We may calibrate the camera quickly by truncating equations (9.1) and (9.2) to $O(r^6)$. Then we may numerically solve for the distortion constants K_1, K_2, K_3, P_1, P_2 using an image reference. For a more robust solution we may use multiple reference images, where the chessboard is considered to be in the plane $Z = 0$. In this instance we may consider a set of N images of a chessboard taken parallel to our camera in our current directory. We may rotate the chessboard in the XY plane in order to have a better measure of the radial distortion points.

It is said that each *object point* is the corner of where two dark squares touch each other in a chessboard. We will detect these points using the `OpenCV` function `cv2.findChessboardCorners`.

We begin the code (see listing 9.1) by loading the images and assuming that the chessboard image (in .jpg format) has $6 \times 7 = 42$ visible *object points*. We may then obtain a system of equations with 42 equations that may be solved numerically in order to obtain K_1, K_2, K_3, P_1, P_2.

The transformation that is found using `OpenCV`'s `cv2.calibrateCamera` returns an intrinsic camera matrix as well as rotation and translation vectors of the form

$$A = \begin{pmatrix} f_x & 0 & x_c \\ 0 & f_y & y_c \\ 0 & 0 & 1 \end{pmatrix}, \tag{9.3}$$

where

$$\begin{pmatrix} u \\ v \\ 1 \end{pmatrix} = \begin{pmatrix} f_x & 0 & x_c \\ 0 & f_y & y_c \\ 0 & 0 & 1 \end{pmatrix} \mathbf{R} \begin{pmatrix} X \\ Y \\ Z \\ 1 \end{pmatrix}, \tag{9.4}$$

where $\mathbf{R}$ is a rotation and translation matrix, f_x, f_y are focal points, $u, v = x_d, y_d$ are image points and our object points are $X, Y, Z = x_u, y_u, 0$.

An iterative solution to (9.3), (9.1) and (9.2) is given by using `OpenCV`'s `cv2.calibrateCamera`. Finally we use `cv2.undistort` to save a calibrated image result (see listing 9.2 for the code). The image to be undistorted is called `left12.jpg` and can be seen in figure 9.1.

Having quickly calibrated our camera focus for a certain plane $Z = 0$ we may then move on to have better accuracy in measuring our objects' dimensions.

```
import numpy as np
import cv2
import glob

# termination criteria
criteria = (cv2.TERM_CRITERIA_EPS + cv2.
    TERM_CRITERIA_MAX_ITER, 30, 0.001)

# prepare object points as  (0,0,0), (1,0,0), (2,0,0), etc
objp = np.zeros((6*7,3), np.float32)
objp[:,:2] = np.mgrid[0:7,0:6].T.reshape(-1,2)

# Arrays to store object points and image points from all
    the images.
objpoints = [] # 3d point in real world space
imgpoints = [] # 2d points in image plane.
images = glob.glob("*.jpg")

for fname in images:
    img = cv2.imread(fname)
    gray = cv2.cvtColor(img, cv2.COLOR_BGR2GRAY)

    # Find the chess board corners
    ret, corners = cv2.findChessboardCorners(gray, (7,6),
        None)

    # If found, add object points, image points (after
        refining them)
    if ret == True:
        objpoints.append(objp)
        corners2 = cv2.cornerSubPix(gray,corners, (11,11),
            (-1,-1), criteria)
        imgpoints.append(corners)

        # Draw and display the corners
        cv2.drawChessboardCorners(img, (7,6), corners2, ret
            )
        cv2.imshow("img", img)
        cv2.waitKey(500)
cv2.destroyAllWindows()
```

Listing 9.1. The application of `calibrateCamera`.

9.1.2 Using camera sensor size or a previous distance

Consider the image in figure 9.2.

By using similarity triangles we may obtain the relation

$$d = f\frac{z_b}{z_a}, \tag{9.5}$$

```
ret, mtx, dist, rvecs, tvecs = cv2.calibrateCamera(
    objpoints, imgpoints, gray.shape[::-1],None,None)


img = cv2.imread("left12.jpg")
h,  w = img.shape[:2]
newcameramtx, roi=cv2.getOptimalNewCameraMatrix(mtx,dist,(w
    ,h),1,(w,h))
dst = cv2.undistort(img, mtx, dist, None, newcameramtx)
# crop the image
x,y,w,h = roi
dst = dst[y:y+h, x:x+w]
cv2.imwrite("calibresult.png",dst)
```

Listing 9.2 Application of calibrate undistorted camera.

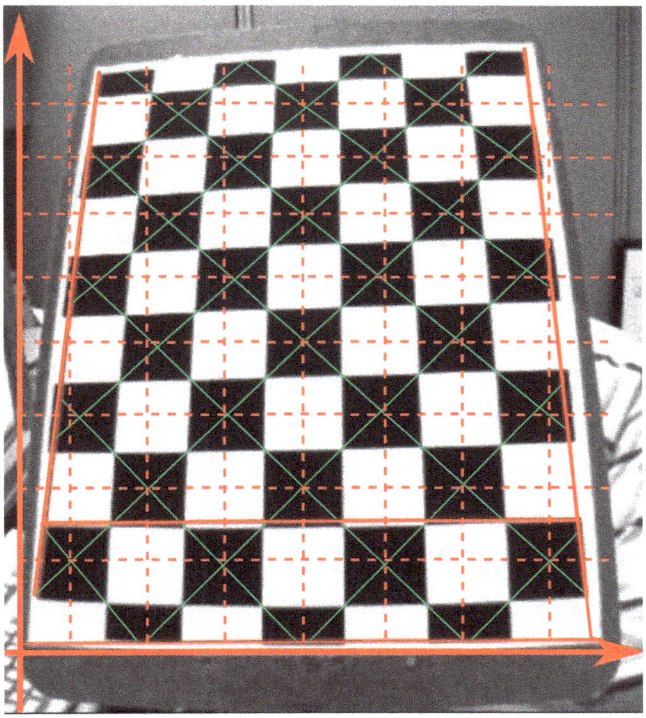

Figure 9.1. Chessboard to calibrate a digital camera.

where $f = \sqrt{f_x^2 + f_y^2}$, z_a is the camera's sensor height in millimetres, z_a is the object's height in millimetres and d is the distance from the camera lens to the object in millimetres.

Thus to obtain the approximate image distance from the camera lens one must know the object's height and know the specifications of the camera's CCD sensor.

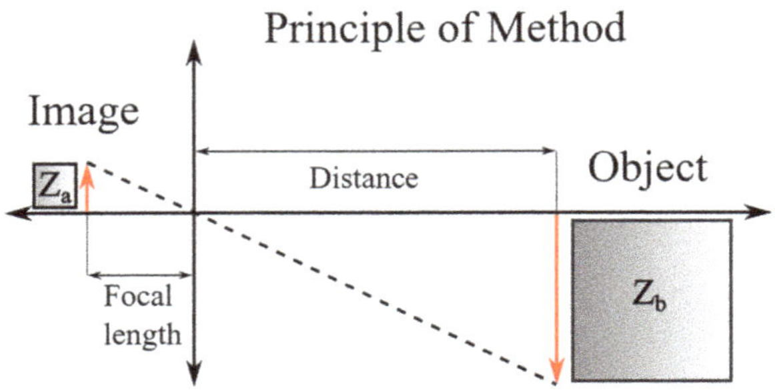

Figure 9.2. Trigonometrical scheme to calibrate a digital camera.

Another way of doing this is to obtain the camera lens focus by using the camera's pixel resolution, a previously known calibration distance and the width of the object. In that manner the only variable to calculate is the camera–object distance.

9.2 Single image depth estimation

Continuing onto depth estimation (Laga *et al* 2020), we may see that there are two main datasets that are common for this task. The NYU-v2 dataset that includes images taken with a monocular camera with their corresponding depth map, and the 2015 KITTI dataset which includes images taken from a monocular camera featuring outside street images and video from a driving car perspective.

Again, the difficulty in implementing deep learning 3D single image depth estimation lies in the dataset quantity and quality. We will quickly review two state-of-the-art depth estimation techniques and share their code implementations.

9.2.1 Consistent video depth estimation

The first implementation for depth estimation is from the paper '*Consistent video depth estimation*' developed by Facebook Research (Luo *et al* 2020), and consists in *geometrically consistent depth*. The term geometric consistency not only implies that the depth maps do not flicker over time but also that all the depth maps are in mutual agreement. That is, the pixels are projected via their depth and camera pose accurately amongst frames. Using this technique a CNN was pretrained for the task and then tuned using each particular video input.

The breakthrough from this technique is consistent in the type of video where most single image depth estimations are only static-based and fail in consistent dynamic video circumstances. For this two geometric losses are computed: (i) spatial loss and (ii) disparity loss. The errors are back-propagated to update the network weights (which are shared across all frames).

To improve pose estimation for videos with dynamic motion, a mask R-CNN was applied to obtain the spatial segmentation of people in an image and remove these regions for more reliable keypoint extraction and matching. This provides accurate intrinsic and extrinsic camera parameters as well as a sparse point cloud reconstruction.

We now describe the geometric loss system. For a given frame pair $(i, j) \in S$, the optical flow field $F_{i\to j}$ describes which pixel pairs show the same scene point. The flow is used to test the geometric consistency of the current depth estimates. If the flow is correct and a flow-displaced point $f_{i\to j}(x)$ is identical to the depth-reprojected point $p_{i\to j}(x)$ then the depth must be consistent.

If x is a 2D pixel coordinate in frame i, the flow-displaced point is

$$f_{i\to j}(x) = x + F_{i\to j}(x). \tag{9.6}$$

We also consider a 3D point $c_i(x)$ in frame i's camera coordinate system. Projecting this point onto frame j we may obtain a transformation of the form

$$c_{i\to j}(x) = R_j^{\mathrm{T}}\left(R_i c_i(x) + \tilde{t}_i - \tilde{t}_j\right). \tag{9.7}$$

Then the spatial loss is given in terms of the pixel position and the flow-displaced point as

$$\mathcal{L}_{i\to j}^{\text{spatial}}(x) = \left\| p_{i\to j}(x) - f_{i\to j}(x) \right\|_2 \tag{9.8}$$

and the disparity loss is given in terms of frame i's focal length u_i and z_i and $z_{i\to j}$ are the scalar z-components of c_i and $c_{i\to j}$, respectively,

$$\mathcal{L}_{i\to j}^{\text{disparity}}(x) = u_i \left| z_{i\to j}^{-1}(x) - z_j^{-1}\left(f_{i\to j}(x)\right) \right| \tag{9.9}$$

so then the total loss may be a linear combination of both of these losses:

$$\mathcal{L}_{i\to j} = \frac{1}{\left|M_{i\to j}\right|} \sum_{x \in M_{i\to j}} \mathcal{L}_{i\to j}^{\text{spatial}}(x) + \lambda \mathcal{L}_{i\to j}^{\text{disparity}}(x), \tag{9.10}$$

where $M_{i\to j}$ is a scaling coefficient.

Implementation

One may use the Google Colab file located at Github https://github.com/facebookresearch/consistent_depth, but for ease of use one may use the code in listing 9.3 in an IPython-Ubuntu setting.

The commands that start with an exclamation mark ! indicate that they are being executed in Bash.

The time for test-time training varies for videos of different lengths. For a video of 244 frames, training on 4 NVIDIA Tesla M40 GPUs takes around 40 min. So the pretrained model and demo are shown here.

```
# -*- coding: utf-8 -*-
"""Consistent Video Depth Estimation.ipynb

Automatically generated by Colaboratory.

Original file is located at
    https://colab.research.google.com/drive/1i5_uVHWOJ1h2adRFT5BuDhoRftq9Oosx

# [SIGGRAPH 2020] Consistent Video Depth Estimation
[Project website](https://github.com/facebookresearch/consistent_depth)

## Download Code
"""

# Commented out IPython magic to ensure Python compatibility.
# %cd /content/
!git clone https://github.com/facebookresearch/consistent_depth.git
# %cd consistent_depth
!git submodule update --init --recursive

"""## Prepare Environment

### Download model files
"""

!./scripts/download_model.sh

"""### Install Python Dependencies"""

!./scripts/install.sh

"""### [Optional] Install COLMAP

> **COLMAP** is not required for running the demo sequence. But if you want to run on your own video and let **COLMAP** to estimate camera pose, please install **COLMAP**. Installation takes a couple minutes.



"""

# Commented out IPython magic to ensure Python compatibility.
# Install packages
```

Listing 9.3 Consistent Video Depth Estimation.ipynb

```
!sudo apt-get install \
    git \
    cmake \
    build-essential \
    libboost-program-options-dev \
    libboost-filesystem-dev \
    libboost-graph-dev \
    libboost-regex-dev \
    libboost-system-dev \
    libboost-test-dev \
    libeigen3-dev \
    libsuitesparse-dev \
    libfreeimage-dev \
    libgoogle-glog-dev \
    libgflags-dev \
    libglew-dev \
    qtbase5-dev \
    libqt5opengl5-dev \
    libcgal-dev \
    libcgal-qt5-dev

# download precompiled colmap packages
# %pushd /content/
!wget https://www.dropbox.com/s/zvbw6h8nrlm3hc3/colmap-packages.zip?dl=1 -O colmap-packages.zip
!unzip colmap-packages.zip -d colmap-packages
# ceres-solver make install
# %cd colmap-packages/
# %pushd ceres-solver/build/
!sudo make install
# %popd
# colmap make install
# %pushd colmap/build
!sudo make install
!CC=/usr/bin/gcc-6 CXX=/usr/bin/g++-6 cmake ..
# %popd
# %popd  # pop /content

# test colmap
!colmap
!which colmap

"""## Prepare Input Video
You can either try out the demo we provied (no need to install COLMAP) or run on your own video sequence (COLMAP required).

### Try out Our Demo
Download the demo video together with its precomputed COLMAP results
```

Listing 9.3 (Continued.)

```
"""

# Commented out IPython magic to ensure Python compatibility.
# %%capture
# !./scripts/download_demo.sh results/ayush
# path = 'results/ayush'
# video_file = 'data/videos/ayush.mp4'
# camera_params = "1671.770118, 540, 960"
# camera_model = "SIMPLE_PINHOLE"

"""### Use Your Own Video
- [Optional] **Calibrate and set camera parameters** using [`PINHOLE` (fx, fy, cx, cy) or `SIMPLE_PINHOLE` (f, cx, cy) model](https://colmap.github.io/cameras.html).
Camera intrinsics calibration is optional but suggested for more accurate and faster camera registration. Just leave the fields with default values if you want to skip this step.
We typically calibrate the camera by capturing a video of a textured plane with really slow camera motion while trying to let target features
cover the full field of view, selecting non-blurry frames, running **COLMAP** on these images.
- **Upload your video**. Short video is more desired. As a reference, our system takes about 37min on a 3s video excluding COLMAP reconstruction part when tested with one NVIDIA GeForce RTX 2080 GPU.
Our system works on videos of static scenes or dynamic scenes containing moderate motion (consistent motion that is not epipolar-aligned or inconsistent motion (e.g., a waving hand)). Consistent motion (e.g., a moving car) can sometimes be aligned with the epipolar geometry and cause our method, like most others, to estimate the wrong depth. As a rule of thumb, large baseline (camera translation) is preferred especially a baseline that is much larger than scene motion. And it is better when the motion is not parallel to the camera motion.
"""

# Upload video.
# %mkdir -p data/videos
# %pushd data/videos

from google.colab import files
import os
from os.path import join as pjoin

uploaded = files.upload()
assert len(uploaded) == 1
for fn in uploaded.keys():
```

Listing 9.3 (Continued.)

```
    print('User uploaded file "{name}" with length {length} bytes'.format(
        name=fn, length=len(uploaded[fn])))

video_file = pjoin('data/videos', list(uploaded.keys())[0])
path = pjoin('results', os.path.splitext(os.path.basename(video_file))[0])
print(f"video file: {video_file}, output path: {path}")
print(f"camera model: {camera_model}, camera parameters: {camera_params}")

# %popd

"""## Run"""

# Commented out IPython magic to ensure Python compatibility.
# %%bash -s "$video_file" "$path" "$camera_model" "$camera_params"
# # convert python variables to bash variables
# video_file="$1"
# path="$2"
# camera_model="$3"
# camera_params="$4"

!echo "Start!"
!python main.py --video_file "$video_file" --path "$path" \
  --camera_params "$camera_params" --camera_model "$camera_model" \
  --make_video
!echo "Done. Your results are saved at $(pwd)/$path."
!echo "Video results are at $(pwd)/$path/R_hierarchical2_mc/videos/."
!echo "Disparity maps are at $(pwd)/$path/R_hierarchical2_mc/B0.1_R1.0_PL1-0_LR0.0004_BS4_Oadam/depth"
```

Listing 9.3 (Continued.)

9.2.2 Adabins

Adabins is a state-of-the-art single image depth estimation algorithm based on the NYU and the KITTI datasets. The main difference is that it does not operate over videos and so the optical flow part of the previous algorithm no longer applies.

This deep learning technique uses a combination of a visual transformer and an encoder–decoder architecture. The encoder–decoder architecture contains a CNN architecture. A transformer-based architecture block is used to divide the depth range into bins whose centre value is estimated adaptively per image. The final depth values are estimated as linear combinations of the bin centres.

The architecture is then two-fold, consisting of two major components: (i) an encoder–decoder block built on a pretrained EfficientNet B5 encoder and a standard feature upsampling decoder and (ii) a proposed adaptive binwidth estimator block called AdaBins.

There are two outputs of the mini-ViT: (i) a vector **b** of bin-widths, which defines how the depth interval D is to be divided for the input image and (ii) range–attention maps that contain useful information for pixel-level depth computation.

One obtains the bin-widths vector **b** as follows:

$$b_i = \frac{b_i' + \epsilon}{\sum_{j=1}^{N}\left(b_j' + \epsilon\right)}, \tag{9.11}$$

where $\epsilon = 10^{-3}$. The small positive ϵ ensures each binwidth is strictly positive. The normalization introduces a competition among the bin-widths and conceptually forces the network to focus on sub-intervals within D by predicting smaller bin-widths at interesting regions of D.

The second part consists in range–attention maps $\mathcal{R}$ passing through a 1×1 convolutional layer to obtain N-channels which is followed by a Softmax activation. The N Softmax scores p_k, $k = 1, \ldots, N$, at each pixel which are probabilities over N depth-bin-centres $c(\mathbf{b}) := \{c(b_1), c(b_2), \ldots, c(b_N)\}$ calculated from bin-widths vector **b** as follows:

$$c(b_i) = d_{\min} + (d_{\max} - d_{\min})\left(b_i/2 + \sum_{j=1}^{i-1} b_j\right). \tag{9.12}$$

Finally, at each pixel the final depth value $\tilde{d}$ is calculated from the linear combination of Softmax scores at that pixel and the depth-bin-centres $c(\mathbf{b})$ as follows:

$$\tilde{d} = \sum_{k=1}^{N} c(b_k)p_k. \tag{9.13}$$

For training, a pixel-wise depth loss is used with a scaled version of the scale-invariant loss (SI) :

$$\mathcal{L}_{\text{pixel}} = \alpha\sqrt{\frac{1}{T}\sum_i g_i^2 - \frac{\lambda}{T^2}\left(\sum_i g_i\right)^2}, \tag{9.14}$$

where $g_i = \log \tilde{d}_i - \log d_i$ with d_i the ground truth depth, and T denotes the number of pixels with valid ground truth values. In the paper by Bhat *et al* (2020) $\lambda = 0.85$ and $\alpha = 10$ were empirically used for all the experiments.

Inference implementation

Make sure you have performed a `pip install` for the following dependences: `torch`, `torchvision`, `tqdm`, `matplotlib`, `Pillow` and `scipy`. Also, this implementation may be seen in the following Github: https://github.com/shariqfarooq123/AdaBins/.

One must also have previously downloaded any of the model paths in order to carry out the inference based on the training sets. These may be saved onto the relative path /pretrained/AdaBins_nyu.pt, for example. The pretrained `models` may be downloaded at https://drive.google.com/drive/folders/1nYyaQXOBjNdUJDsmJpcRpu6oE55aQoLA.

To run the inference on a CPU only, one must change the code of the `__init__` function contained in the InferenceHelper class in `infer.py`. The `device` argument `cuda:0` must be changed to `cpu`. The code in listing 9.4 corresponds to `infer.py`, which we may run on any NYU test image or similar (the hardcoded image path at line 155 may be changed for the image of choice).

The `infer.py` file depends on `model_io.py` which helps us load the weights onto the model and is coded as shown in listing 9.5.

```
1  import glob
2  import os
3
4  import numpy as np
5  import torch
6  import torch.nn as nn
7  from PIL import Image
8  from torchvision import transforms
9  from tqdm import tqdm
10
11 import model_io
12 import utils
13 from models import UnetAdaptiveBins
14
15
16 def _is_pil_image(img):
17     return isinstance(img, Image.Image)
18
19
20 def _is_numpy_image(img):
21     return isinstance(img, np.ndarray) and (img.ndim in {2,
           3})
22
23
24 class ToTensor(object):
25     def __init__(self):
26         self.normalize = transforms.Normalize(mean=[0.485,
             0.456, 0.406], std=[0.229, 0.224, 0.225])
27
28     def __call__(self, image, target_size=(640, 480)):
29         # image = image.resize(target_size)
30         image = self.to_tensor(image)
31         image = self.normalize(image)
32         return image
```

Listing 9.4 `infer.py`

```python

    def to_tensor(self, pic):
        if not (_is_pil_image(pic) or _is_numpy_image(pic)):
            raise TypeError(
                'pic␣should␣be␣PIL␣Image␣or␣ndarray.␣Got␣{}'.format(type(pic)))

        if isinstance(pic, np.ndarray):
            img = torch.from_numpy(pic.transpose((2, 0, 1)))
            return img

        # handle PIL Image
        if pic.mode == 'I':
            img = torch.from_numpy(np.array(pic, np.int32, copy=False))
        elif pic.mode == 'I;16':
            img = torch.from_numpy(np.array(pic, np.int16, copy=False))
        else:
            img = torch.ByteTensor(torch.ByteStorage.from_buffer(pic.tobytes()))
        # PIL image mode: 1, L, P, I, F, RGB, YCbCr, RGBA, CMYK
        if pic.mode == 'YCbCr':
            nchannel = 3
        elif pic.mode == 'I;16':
            nchannel = 1
        else:
            nchannel = len(pic.mode)
        img = img.view(pic.size[1], pic.size[0], nchannel)

        img = img.transpose(0, 1).transpose(0, 2).contiguous()
        if isinstance(img, torch.ByteTensor):
            return img.float()
        else:
            return img


class InferenceHelper:
    def __init__(self, dataset='nyu', device='cuda:0'):
        self.toTensor = ToTensor()
        self.device = device
        if dataset == 'nyu':
            self.min_depth = 1e-3
            self.max_depth = 10
            self.saving_factor = 1000  # used to save in 16 bit
```

Listing 9.4 (Continued.)

```
            model = UnetAdaptiveBins.build(n_bins=256, min_val=self.min_depth, max_val=self.max_depth)
            pretrained_path = "./pretrained/AdaBins_nyu.pt"
        elif dataset == 'kitti':
            self.min_depth = 1e-3
            self.max_depth = 80
            self.saving_factor = 256
            model = UnetAdaptiveBins.build(n_bins=256, min_val=self.min_depth, max_val=self.max_depth)
            pretrained_path = "./pretrained/AdaBins_kitti.pt"
        else:
            raise ValueError("dataset can be either 'nyu' or 'kitti' but got {}".format(dataset))

        model, _, _ = model_io.load_checkpoint(pretrained_path, model)
        model.eval()
        self.model = model.to(self.device)

    @torch.no_grad()
    def predict_pil(self, pil_image, visualized=False):
        # pil_image = pil_image.resize((640, 480))
        img = np.asarray(pil_image) / 255.

        img = self.toTensor(img).unsqueeze(0).float().to(self.device)
        bin_centers, pred = self.predict(img)

        if visualized:
            viz = utils.colorize(torch.from_numpy(pred).unsqueeze(0), vmin=None, vmax=None, cmap='magma')
            # pred = np.asarray(pred*1000, dtype='uint16')
            viz = Image.fromarray(viz)
            return bin_centers, pred, viz
        return bin_centers, pred

    @torch.no_grad()
    def predict(self, image):
        bins, pred = self.model(image)
        pred = np.clip(pred.cpu().numpy(), self.min_depth, self.max_depth)

        # Flip
        image = torch.Tensor(np.array(image.cpu().numpy())[..., ::-1].copy()).to(self.device)
        pred_lr = self.model(image)[-1]
```

Listing 9.4 (Continued.)

```
            pred_lr = np.clip(pred_lr.cpu().numpy()[..., ::-1], self.min_depth, self.max_depth)

            # Take average of original and mirror
            final = 0.5 * (pred + pred_lr)
            final = nn.functional.interpolate(torch.Tensor(final), image.shape[-2:],
                                              mode='bilinear', align_corners=True).cpu().numpy()

            final[final < self.min_depth] = self.min_depth
            final[final > self.max_depth] = self.max_depth
            final[np.isinf(final)] = self.max_depth
            final[np.isnan(final)] = self.min_depth

            centers = 0.5 * (bins[:, 1:] + bins[:, :-1])
            centers = centers.cpu().squeeze().numpy()
            centers = centers[centers > self.min_depth]
            centers = centers[centers < self.max_depth]

            return centers, final

        @torch.no_grad()
        def predict_dir(self, test_dir, out_dir):
            os.makedirs(out_dir, exist_ok=True)
            transform = ToTensor()
            all_files = glob.glob(os.path.join(test_dir, "*"))
            self.model.eval()
            for f in tqdm(all_files):
                image = np.asarray(Image.open(f), dtype='float32') / 255.
                image = transform(image).unsqueeze(0).to(self.device)

                centers, final = self.predict(image)
                # final = final.squeeze().cpu().numpy()

                final = (final * self.saving_factor).astype('uint16')
                basename = os.path.basename(f).split('.')[0]
                save_path = os.path.join(out_dir, basename + ".png")

                Image.fromarray(final).save(save_path)


    if __name__ == '__main__':
        import matplotlib.pyplot as plt
        from time import time
```

Listing 9.4 (Continued.)

```
    img = Image.open("test_imgs/classroom__rgb_00283.jpg") #testing img
    start = time()
    inferHelper = InferenceHelper()
    centers, pred = inferHelper.predict_pil(img)
    print(f"took :{time() - start}s")
    plt.imshow(pred.squeeze(), cmap='magma_r')
    plt.show()
```

Listing 9.4 (Continued.)

```
import os
import torch

def save_weights(model, filename, path="./saved_models"):
    if not os.path.isdir(path):
        os.makedirs(path)

    fpath = os.path.join(path, filename)
    torch.save(model.state_dict(), fpath)
    return

def save_checkpoint(model, optimizer, epoch, filename, root="./checkpoints"):
    if not os.path.isdir(root):
        os.makedirs(root)

    fpath = os.path.join(root, filename)
    torch.save(
        {
            "model": model.state_dict(),
            "optimizer": optimizer.state_dict(),
            "epoch": epoch
        }
        , fpath)

def load_weights(model, filename, path="./saved_models"):
    fpath = os.path.join(path, filename)
    state_dict = torch.load(fpath)
    model.load_state_dict(state_dict)
    return model

def load_checkpoint(fpath, model, optimizer=None):
    ckpt = torch.load(fpath, map_location='cpu')
    if optimizer is None:
        optimizer = ckpt.get('optimizer', None)
    else:
```

Listing 9.5 `model_io.py`

```
            optimizer.load_state_dict(ckpt['optimizer'])
        epoch = ckpt['epoch']

        if 'model' in ckpt:
            ckpt = ckpt['model']
        load_dict = {}
        for k, v in ckpt.items():
            if k.startswith('module.'):
                k_ = k.replace('module.', '')
                load_dict[k_] = v
            else:
                load_dict[k] = v

        modified = {}  # backward compatibility to older naming
            of architecture blocks
        for k, v in load_dict.items():
            if k.startswith('adaptive_bins_layer.embedding_conv
                .'):
                k_ = k.replace('adaptive_bins_layer.
                    embedding_conv.',
                               'adaptive_bins_layer.conv3x3.')
                modified[k_] = v
                # del load_dict[k]

            elif k.startswith('adaptive_bins_layer.
                patch_transformer.embedding_encoder'):

                k_ = k.replace('adaptive_bins_layer.
                    patch_transformer.embedding_encoder',
                               'adaptive_bins_layer.
                                   patch_transformer.
                                   embedding_convPxP')
                modified[k_] = v
                # del load_dict[k]
            else:
                modified[k] = v  # else keep the original

        model.load_state_dict(modified)
        return model, optimizer, epoch
```

Listing 9.5 (Continued.)

The previous .py file uses an `utils.py` file to work, which is seen in the code in listing 9.6.

We must also consider a `models` folder in our relative path, which contains three key pieces for the model: `unet_adaptive_bins.py` (listing 9.7), `miniViT.py` (listing 9.8) and `layers.py` (listing 9.9).

```
import base64
import math
import re
from io import BytesIO

import matplotlib.cm
import numpy as np
import torch
import torch.nn
from PIL import Image


class RunningAverage:
    def __init__(self):
        self.avg = 0
        self.count = 0

    def append(self, value):
        self.avg = (value + self.count * self.avg) / (self.
            count + 1)
        self.count += 1

    def get_value(self):
        return self.avg


def denormalize(x, device='cpu'):
    mean = torch.Tensor([0.485, 0.456, 0.406]).view(1, 3,
        1, 1).to(device)
    std = torch.Tensor([0.229, 0.224, 0.225]).view(1, 3, 1,
        1).to(device)
    return x * std + mean


class RunningAverageDict:
    def __init__(self):
        self._dict = None

    def update(self, new_dict):
        if self._dict is None:
            self._dict = dict()
            for key, value in new_dict.items():
                self._dict[key] = RunningAverage()

        for key, value in new_dict.items():
            self._dict[key].append(value)

    def get_value(self):
        return {key: value.get_value() for key, value in
            self._dict.items()}
```

Listing 9.6 uitls.py

```


def colorize(value, vmin=10, vmax=1000, cmap='magma_r'):
    value = value.cpu().numpy()[0, :, :]
    invalid_mask = value == -1

    # normalize
    vmin = value.min() if vmin is None else vmin
    vmax = value.max() if vmax is None else vmax

    if vmin != vmax:
        value = (value - vmin) / (vmax - vmin)  # vmin..vmax

    else:
        # Avoid 0-division
        value = value * 0.

    # squeeze last dim if it exists
    # value = value.squeeze(axis=0)

    cmapper = matplotlib.cm.get_cmap(cmap)
    value = cmapper(value, bytes=True)  # (nxmx4)
    value[invalid_mask] = 255
    img = value[:, :, :3]

    #     return img.transpose((2, 0, 1))
    return img

def count_parameters(model):
    return sum(p.numel() for p in model.parameters() if p.requires_grad)

def compute_errors(gt, pred):
    thresh = np.maximum((gt / pred), (pred / gt))

    a1 = (thresh < 1.25).mean()
    a2 = (thresh < 1.25 ** 2).mean()
    a3 = (thresh < 1.25 ** 3).mean()

    abs_rel = np.mean(np.abs(gt - pred) / gt)
    sq_rel = np.mean(((gt - pred) ** 2) / gt)

    rmse = (gt - pred) ** 2
    rmse = np.sqrt(rmse.mean())

    rmse_log = (np.log(gt) - np.log(pred)) ** 2
    rmse_log = np.sqrt(rmse_log.mean())

    err = np.log(pred) - np.log(gt)
    silog = np.sqrt(np.mean(err ** 2) - np.mean(err) ** 2) * 100
```

Listing 9.6 (Continued.)

```
    log_10 = (np.abs(np.log10(gt) - np.log10(pred))).mean()

    return dict(a1=a1, a2=a2, a3=a3, abs_rel=abs_rel, rmse=rmse, log_10=log_10, rmse_log=rmse_log,
                silog=silog, sq_rel=sq_rel)

##################################### Demo Utilities ############################################

def b64_to_pil(b64string):
    image_data = re.sub('^data:image/.+;base64,', '', b64string)
    # image = Image.open(cStringIO.StringIO(image_data))
    return Image.open(BytesIO(base64.b64decode(image_data)))

# Compute edge magnitudes
from scipy import ndimage

def edges(d):
    dx = ndimage.sobel(d, 0)  # horizontal derivative
    dy = ndimage.sobel(d, 1)  # vertical derivative
    return np.abs(dx) + np.abs(dy)

class PointCloudHelper():
    def __init__(self, width=640, height=480):
        self.xx, self.yy = self.worldCoords(width, height)

    def worldCoords(self, width=640, height=480):
        hfov_degrees, vfov_degrees = 57, 43
        hFov = math.radians(hfov_degrees)
        vFov = math.radians(vfov_degrees)
        cx, cy = width / 2, height / 2
        fx = width / (2 * math.tan(hFov / 2))
        fy = height / (2 * math.tan(vFov / 2))
        xx, yy = np.tile(range(width), height), np.repeat(range(height), width)
        xx = (xx - cx) / fx
        yy = (yy - cy) / fy
        return xx, yy

    def depth_to_points(self, depth):
        depth[edges(depth) > 0.3] = np.nan  # Hide depth edges
        length = depth.shape[0] * depth.shape[1]
        # depth[edges(depth) > 0.3] = 1e6  # Hide depth edges
        z = depth.reshape(length)

        return np.dstack((self.xx * z, self.yy * z, z)).reshape((length, 3))
```

Listing 9.6 (Continued.)

```python
import torch
import torch.nn as nn
import torch.nn.functional as F
import torch.nn.functional as F

from .miniViT import mViT


class UpSampleBN(nn.Module):
    def __init__(self, skip_input, output_features):
        super(UpSampleBN, self).__init__()

        self._net = nn.Sequential(nn.Conv2d(skip_input, output_features, kernel_size=3, stride=1, padding=1),
                                  nn.BatchNorm2d(output_features),
                                  nn.LeakyReLU(),
                                  nn.Conv2d(output_features, output_features, kernel_size=3, stride=1, padding=1),
                                  nn.BatchNorm2d(output_features),
                                  nn.LeakyReLU())

    def forward(self, x, concat_with):
        up_x = F.interpolate(x, size=[concat_with.size(2), concat_with.size(3)], mode='bilinear', align_corners=True)
        f = torch.cat([up_x, concat_with], dim=1)
        return self._net(f)


class DecoderBN(nn.Module):
    def __init__(self, num_features=2048, num_classes=1, bottleneck_features=2048):
        super(DecoderBN, self).__init__()
        features = int(num_features)

        self.conv2 = nn.Conv2d(bottleneck_features, features, kernel_size=1, stride=1, padding=1)

        self.up1 = UpSampleBN(skip_input=features // 1 + 112 + 64, output_features=features // 2)
        self.up2 = UpSampleBN(skip_input=features // 2 + 40 + 24, output_features=features // 4)
        self.up3 = UpSampleBN(skip_input=features // 4 + 24 + 16, output_features=features // 8)
        self.up4 = UpSampleBN(skip_input=features // 8 + 16 + 8, output_features=features // 16)
```

Listing 9.7. `unet_adaptive_bins.py`

```python

        #         self.up5 = UpSample(skip_input=features // 16 + 3, output_features=features//16)
        self.conv3 = nn.Conv2d(features // 16, num_classes, kernel_size=3, stride=1, padding=1)
        # self.act_out = nn.Softmax(dim=1) if output_activation == 'softmax' else nn.Identity()

    def forward(self, features):
        x_block0, x_block1, x_block2, x_block3, x_block4 = features[4], features[5], features[6], features[8], features[
            11]

        x_d0 = self.conv2(x_block4)

        x_d1 = self.up1(x_d0, x_block3)
        x_d2 = self.up2(x_d1, x_block2)
        x_d3 = self.up3(x_d2, x_block1)
        x_d4 = self.up4(x_d3, x_block0)
        #         x_d5 = self.up5(x_d4, features[0])
        out = self.conv3(x_d4)
        # out = self.act_out(out)
        # if with_features:
        #     return out, features[-1]
        # elif with_intermediate:
        #     return out, [x_block0, x_block1, x_block2, x_block3, x_block4, x_d1, x_d2, x_d3, x_d4]
        return out


class Encoder(nn.Module):
    def __init__(self, backend):
        super(Encoder, self).__init__()
        self.original_model = backend

    def forward(self, x):
        features = [x]
        for k, v in self.original_model._modules.items():
            if (k == 'blocks'):
                for ki, vi in v._modules.items():
                    features.append(vi(features[-1]))
            else:
                features.append(v(features[-1]))
        return features


class UnetAdaptiveBins(nn.Module):
    def __init__(self, backend, n_bins=100, min_val=0.1, max_val=10, norm='linear'):
```

Listing 9.7. (Continued.)

```
        super(UnetAdaptiveBins, self).__init__()
        self.num_classes = n_bins
        self.min_val = min_val
        self.max_val = max_val
        self.encoder = Encoder(backend)
        self.adaptive_bins_layer = mViT(128, n_query_channels=128, patch_size=16,
                                        dim_out=n_bins,
                                        embedding_dim=128, norm=norm)

        self.decoder = DecoderBN(num_classes=128)
        self.conv_out = nn.Sequential(nn.Conv2d(128, n_bins, kernel_size=1, stride=1, padding=0),
                                      nn.Softmax(dim=1))

    def forward(self, x, **kwargs):
        unet_out = self.decoder(self.encoder(x), **kwargs)
        bin_widths_normed, range_attention_maps = self.adaptive_bins_layer(unet_out)
        out = self.conv_out(range_attention_maps)

        # Post process
        # n, c, h, w = out.shape
        # hist = torch.sum(out.view(n, c, h * w), dim=2) / (h * w)  # not used for training

        bin_widths = (self.max_val - self.min_val) * bin_widths_normed  # .shape = N, dim_out
        bin_widths = nn.functional.pad(bin_widths, (1, 0), mode='constant', value=self.min_val)
        bin_edges = torch.cumsum(bin_widths, dim=1)

        centers = 0.5 * (bin_edges[:, :-1] + bin_edges[:, 1:])
        n, dout = centers.size()
        centers = centers.view(n, dout, 1, 1)

        pred = torch.sum(out * centers, dim=1, keepdim=True)

        return bin_edges, pred

    def get_1x_lr_params(self):  # lr/10 learning rate
        return self.encoder.parameters()

    def get_10x_lr_params(self):  # lr learning rate
        modules = [self.decoder, self.adaptive_bins_layer, self.conv_out]
        for m in modules:
```

Listing 9.7. (Continued.)

```
                yield from m.parameters()

    @classmethod
    def build(cls, n_bins, **kwargs):
        basemodel_name = 'tf_efficientnet_b5_ap'

        print('Loading base model ()...'.format(basemodel_name), end='')
        basemodel = torch.hub.load('rwightman/gen-efficientnet-pytorch', basemodel_name, pretrained=True)
        print('Done.')

        # Remove last layer
        print('Removing last two layers (global_pool & classifier).')
        basemodel.global_pool = nn.Identity()
        basemodel.classifier = nn.Identity()

        # Building Encoder-Decoder model
        print('Building Encoder-Decoder model..', end='')
        m = cls(basemodel, n_bins=n_bins, **kwargs)
        print('Done.')
        return m


if __name__ == '__main__':
    model = UnetAdaptiveBins.build(100)
    x = torch.rand(2, 3, 480, 640)
    bins, pred = model(x)
    print(bins.shape, pred.shape)
```

Listing 9.7. (Continued.)

```
import torch
import torch.nn as nn

from .layers import PatchTransformerEncoder, PixelWiseDotProduct


class mViT(nn.Module):
    def __init__(self, in_channels, n_query_channels=128, patch_size=16, dim_out=256,
                 embedding_dim=128, num_heads=4, norm='linear'):
        super(mViT, self).__init__()
```

Listing 9.8 `miniViT.py`

```
        self.norm = norm
        self.n_query_channels = n_query_channels
        self.patch_transformer = PatchTransformerEncoder(in_channels, patch_size, embedding_dim, num_heads)
        self.dot_product_layer = PixelWiseDotProduct()

        self.conv3x3 = nn.Conv2d(in_channels, embedding_dim, kernel_size=3, stride=1, padding=1)
        self.regressor = nn.Sequential(nn.Linear(embedding_dim, 256),
                                       nn.LeakyReLU(),
                                       nn.Linear(256, 256),
                                       nn.LeakyReLU(),
                                       nn.Linear(256, dim_out))

    def forward(self, x):
        # n, c, h, w = x.size()
        tgt = self.patch_transformer(x.clone())  # .shape = S, N, E

        x = self.conv3x3(x)

        regression_head, queries = tgt[0, ...], tgt[1:self.n_query_channels + 1, ...]

        # Change from S, N, E to N, S, E
        queries = queries.permute(1, 0, 2)
        range_attention_maps = self.dot_product_layer(x, queries)  # .shape = n, n_query_channels, h, w

        y = self.regressor(regression_head)  # .shape = N, dim_out
        if self.norm == 'linear':
            y = torch.relu(y)
            eps = 0.1
            y = y + eps
        elif self.norm == 'softmax':
            return torch.softmax(y, dim=1), range_attention_maps
        else:
            y = torch.sigmoid(y)
        y = y / y.sum(dim=1, keepdim=True)
        return y, range_attention_maps
```

Listing 9.8 (Continued.)

9.3 State-of-the-art real-time facial detection

9.3.1 Introduction

Since 2015 the state-of-the-art accuracy in `models` using the dataset Labelled Faces in the Wild (LFW) has not increased much. Developed by Google, the major breakthrough was through the use of a CNN algorithm whose architecture is called FaceNet (Schroff *et al* 2015).

In this section we will use FaceNet, which is a state-of-the-art LFW dataset. FaceNet directly learns a mapping from face images to a compact Euclidean space where distances directly correspond to a measure of face similarity. Once this space has been produced, tasks such as facial recognition, verification and clustering can be easily implemented using standard techniques with FaceNet embeddings as feature vectors. The method uses a deep convolutional network trained to directly optimize the embedding.

Another method, ArcFace, was initially described in an arXiv technical report (Deng *et al* 2019). By using this module one can simply achieve LFW 99.83%+ and Megaface 98%+ with a single model.

```python
import torch
import torch.nn as nn


class PatchTransformerEncoder(nn.Module):
    def __init__(self, in_channels, patch_size=10, embedding_dim=128, num_heads=4):
        super(PatchTransformerEncoder, self).__init__()
        encoder_layers = nn.TransformerEncoderLayer(embedding_dim, num_heads, dim_feedforward=1024)
        self.transformer_encoder = nn.TransformerEncoder(encoder_layers, num_layers=4)  # takes shape S,N,E

        self.embedding_convPxP = nn.Conv2d(in_channels, embedding_dim,
                                           kernel_size=patch_size, stride=patch_size, padding=0)

        self.positional_encodings = nn.Parameter(torch.rand(500, embedding_dim), requires_grad=True)

    def forward(self, x):
        embeddings = self.embedding_convPxP(x).flatten(2)  # .shape = n,c,s = n, embedding_dim, s
        # embeddings = nn.functional.pad(embeddings, (1,0))  # extra special token at start ?
```

Listing 9.9 `layers.py`

```python
import torch
import torch.nn as nn


class PatchTransformerEncoder(nn.Module):
    def __init__(self, in_channels, patch_size=10, embedding_dim=128, num_heads=4):
        super(PatchTransformerEncoder, self).__init__()
        encoder_layers = nn.TransformerEncoderLayer(embedding_dim, num_heads, dim_feedforward=1024)
        self.transformer_encoder = nn.TransformerEncoder(encoder_layers, num_layers=4)  # takes shape S,N,E

        self.embedding_convPxP = nn.Conv2d(in_channels, embedding_dim,
                                           kernel_size=patch_size, stride=patch_size, padding=0)

        self.positional_encodings = nn.Parameter(torch.rand(500, embedding_dim), requires_grad=True)

    def forward(self, x):
        embeddings = self.embedding_convPxP(x).flatten(2)  # .shape = n,c,s = n, embedding_dim, s
        # embeddings = nn.functional.pad(embeddings, (1,0))  # extra special token at start ?
```

Listing 9.9 (Continued.)

Facial extraction implementation

We also use the multitask cascaded convolutional neural network (MTCNN) for face detection, e.g. finding and extracting faces from photos. This is a state-of-the-art deep learning model for face detection, described in the 2016 paper '*Joint face detection and alignment using multitask cascaded convolutional networks*' (Zhang *et al* 2016).

We will use the implementation provided by Iván de Paz Centeno in the ipazc/mtcnn project. This can also be installed via pip as per listing 9.10.

We must first download the pretrained model weights trained with Keras, as per listing 9.11.

From here we work with MTCNN to automatically detect faces where possible, and have training and test samples (see listing 9.12 for the code).

FaceNet implementation

The code for the FaceNet implementation is given in listing 9.13.

```
!pip install mtcnn
```

Listing 9.10 Application of calibrate collaboration project.

```
from keras.models import load_model
# load the model
model = load_model('facenet_keras.h5')
```

Listing 9.11 Application of calibrate collaboration project.

```

# function for face detection with mtcnn
from PIL import Image
from numpy import asarray
from mtcnn.mtcnn import MTCNN

# extract a single face from a given photograph
def extract_face(filename, required_size=(160, 160)):
        # load image from file
        image = Image.open(filename)
        # convert to RGB, if needed
        image = image.convert('RGB')
        # convert to array
        pixels = asarray(image)
        # create the detector, using default weights
        detector = MTCNN()
        # detect faces in the image
        results = detector.detect_faces(pixels)
        # extract the bounding box from the first face
        x1, y1, width, height = results[0]['box']
        # bug fix
        x1, y1 = abs(x1), abs(y1)
        x2, y2 = x1 + width, y1 + height
        # extract the face
        face = pixels[y1:y2, x1:x2]
        # resize pixels to the model size
        image = Image.fromarray(face)
        image = image.resize(required_size)
        face_array = asarray(image)
        return face_array

# load the photo and extract the face
pixels = extract_face('img.jpg')
```

Listing 9.12 Application of calibrate collaboration project.

```python

from os import listdir
from PIL import Image
from numpy import asarray
from matplotlib import pyplot
from mtcnn.mtcnn import MTCNN

# extract a single face from a given photograph
def extract_face(filename, required_size=(160, 160)):
        # load image from file
        image = Image.open(filename)
        # convert to RGB, if needed
        image = image.convert('RGB')
        # convert to array
        pixels = asarray(image)
        # create the detector, using default weights
        detector = MTCNN()
        # detect faces in the image
        results = detector.detect_faces(pixels)
        # extract the bounding box from the first face
        x1, y1, width, height = results[0]['box']
        # bug fix
        x1, y1 = abs(x1), abs(y1)
        x2, y2 = x1 + width, y1 + height
        # extract the face
        face = pixels[y1:y2, x1:x2]
        # resize pixels to the model size
        image = Image.fromarray(face)
        image = image.resize(required_size)
        face_array = asarray(image)
        return face_array

# specify folder to plot
folder = '5-celebrity-faces-dataset/train/ben_afflek/'
i = 1
# enumerate files
for filename in listdir(folder):
        # path
        path = folder + filename
        # get face
        face = extract_face(path)
        print(i, face.shape)
        # plot
        pyplot.subplot(2, 7, i)
        pyplot.axis('off')
        pyplot.imshow(face)
        i += 1
pyplot.show()


```

Listing 9.13 Application of calibrate collaboration project.

```
# face detection for the 5 Celebrity Faces Dataset
from os import listdir
from os.path import isdir
from PIL import Image
from matplotlib import pyplot
from numpy import savez_compressed
from numpy import asarray
from mtcnn.mtcnn import MTCNN

# extract a single face from a given photograph
def extract_face(filename, required_size=(160, 160)):
    # load image from file
    image = Image.open(filename)
    # convert to RGB, if needed
    image = image.convert('RGB')
    # convert to array
    pixels = asarray(image)
    # create the detector, using default weights
    detector = MTCNN()
    # detect faces in the image
    results = detector.detect_faces(pixels)
    # extract the bounding box from the first face
    x1, y1, width, height = results[0]['box']
    # bug fix
    x1, y1 = abs(x1), abs(y1)
    x2, y2 = x1 + width, y1 + height
    # extract the face
    face = pixels[y1:y2, x1:x2]
    # resize pixels to the model size
    image = Image.fromarray(face)
    image = image.resize(required_size)
    face_array = asarray(image)
    return face_array

# load images and extract faces for all images in a directory
def load_faces(directory):
    faces = list()
    # enumerate files
    for filename in listdir(directory):
        # path
        path = directory + filename
        # get face
        face = extract_face(path)
        # store
        faces.append(face)
    return faces

# load a dataset that contains one subdir for each class that in turn contains images
```

Listing 9.13 (Continued.)

```
def load_dataset(directory):
        X, y = list(), list()
        # enumerate folders, on per class
        for subdir in listdir(directory):
                # path
                path = directory + subdir + '/'
                # skip any files that might be in the dir
                if not isdir(path):
                        continue
                # load all faces in the subdirectory
                faces = load_faces(path)
                # create labels
                labels = [subdir for _ in range(len(faces))]
                # summarize progress
                print('>loaded %d examples for class: %s' % (len(faces), subdir))
                # store
                X.extend(faces)
                y.extend(labels)
        return asarray(X), asarray(y)

# load train dataset
trainX, trainy = load_dataset('5-celebrity-faces-dataset/train/')
print(trainX.shape, trainy.shape)
# load test dataset
testX, testy = load_dataset('5-celebrity-faces-dataset/val/')
# save arrays to one file in compressed format
savez_compressed('5-celebrity-faces-dataset.npz', trainX, trainy, testX, testy)
```

Listing 9.13 (Continued.)

Facial recognition

We then use a linear SVM to calculate the difference between every person (see listings 9.14 and 9.15).

With example plots given in listing 9.16.

9.4 Fruit classification

For this problem that may frequently arise during common OK and not-OK simple inspection processes, we may want to use binary or multi-class classification. For real-time application inference we also want to use a lightweight model with fewer parameters than, say, a Resnet-50 architecture (Sandler *et al* 2018). For these types of tasks, SSDs such as YOLOv4 or MobileNetv2 may be used (Howard *et al* 2017, Kleiman and Page 2019).

In this instance we train a multi-class classification dataset using Google Colab. The fruit dataset may be collected via this Google Drive link: https://drive.google.com/drive/folders/1FZbG_9PPpRVTNuGPY-XudqWvIRmmuqiE?usp=sharing. Everything

```python

# calculate a face embedding for each face in the dataset using facenet
from numpy import load
from numpy import expand_dims
from numpy import asarray
from numpy import savez_compressed
from keras.models import load_model

# get the face embedding for one face
def get_embedding(model, face_pixels):
        # scale pixel values
        face_pixels = face_pixels.astype('float32')
        # standardize pixel values across channels (global)
        mean, std = face_pixels.mean(), face_pixels.std()
        face_pixels = (face_pixels - mean) / std
        # transform face into one sample
        samples = expand_dims(face_pixels, axis=0)
        # make prediction to get embedding
        yhat = model.predict(samples)
        return yhat[0]

# load the face dataset
data = load('5-celebrity-faces-dataset.npz')
trainX, trainy, testX, testy = data['arr_0'], data['arr_1'], data['arr_2'], data['arr_3']
print('Loaded: ', trainX.shape, trainy.shape, testX.shape, testy.shape)
# load the facenet model
model = load_model('facenet_keras.h5')
print('Loaded Model')
# convert each face in the train set to an embedding
newTrainX = list()
for face_pixels in trainX:
        embedding = get_embedding(model, face_pixels)
        newTrainX.append(embedding)
newTrainX = asarray(newTrainX)
print(newTrainX.shape)
# convert each face in the test set to an embedding
newTestX = list()
for face_pixels in testX:
        embedding = get_embedding(model, face_pixels)
        newTestX.append(embedding)
newTestX = asarray(newTestX)
print(newTestX.shape)
# save arrays to one file in compressed format
savez_compressed('5-celebrity-faces-embeddings.npz', newTrainX, trainy, newTestX, testy)
```

Listing 9.14 Application of calibrate collaboration project.

```python

# develop a classifier for the 5 Celebrity Faces Dataset
from numpy import load
from sklearn.metrics import accuracy_score
from sklearn.preprocessing import LabelEncoder
from sklearn.preprocessing import Normalizer
from sklearn.svm import SVC
# load dataset
data = load('5-celebrity-faces-embeddings.npz')
trainX, trainy, testX, testy = data['arr_0'], data['arr_1'], data['arr_2'], data['arr_3']
print('Dataset: train=%d, test=%d' % (trainX.shape[0], testX.shape[0]))
# normalize input vectors
in_encoder = Normalizer(norm='l2')
trainX = in_encoder.transform(trainX)
testX = in_encoder.transform(testX)
# label encode targets
out_encoder = LabelEncoder()
out_encoder.fit(trainy)
trainy = out_encoder.transform(trainy)
testy = out_encoder.transform(testy)
# fit model
model = SVC(kernel='linear', probability=True)
model.fit(trainX, trainy)
# predict
yhat_train = model.predict(trainX)
yhat_test = model.predict(testX)
# score
score_train = accuracy_score(trainy, yhat_train)
score_test = accuracy_score(testy, yhat_test)
# summarize
print('Accuracy: train=%.3f, test=%.3f' % (score_train*100, score_test*100))
```

Listing 9.15 Application of calibrate collaboration project.

```python
# develop a classifier for the 5 Celebrity Faces Dataset
from random import choice
from numpy import load
from numpy import expand_dims
from sklearn.preprocessing import LabelEncoder
from sklearn.preprocessing import Normalizer
from sklearn.svm import SVC
from matplotlib import pyplot

# load faces
data = load('5-celebrity-faces-dataset.npz')
testX_faces = data['arr_2']
```

Listing 9.16 Application of calibrate collaboration project.

```

# load face embeddings
data = load('5-celebrity-faces-embeddings.npz')
trainX, trainy, testX, testy = data['arr_0'], data['arr_1'], data['arr_2'], data['arr_3']

# normalize input vectors
in_encoder = Normalizer(norm='l2')
trainX = in_encoder.transform(trainX)
testX = in_encoder.transform(testX)

# label encode targets
out_encoder = LabelEncoder()
out_encoder.fit(trainy)
trainy = out_encoder.transform(trainy)
testy = out_encoder.transform(testy)

# fit model
model = SVC(kernel='linear', probability=True)
model.fit(trainX, trainy)

# test model on a random example from the test dataset
selection = choice([i for i in range(testX.shape[0])])
random_face_pixels = testX_faces[selection]
random_face_emb = testX[selection]
random_face_class = testy[selection]
random_face_name = out_encoder.inverse_transform([random_face_class])

# prediction for the face
samples = expand_dims(random_face_emb, axis=0)
yhat_class = model.predict(samples)
yhat_prob = model.predict_proba(samples)

# get name
class_index = yhat_class[0]
class_probability = yhat_prob[0,class_index] * 100
predict_names = out_encoder.inverse_transform(yhat_class)
print('Predicted: %s (%.3f)' % (predict_names[0], class_probability))
print('Expected: %s' % random_face_name[0])

# plot for fun
pyplot.imshow(random_face_pixels)
title = '%s (%.3f)' % (predict_names[0], class_probability)
pyplot.title(title)
pyplot.show()
```

Listing 9.16 (Continued.)

```
# -*- coding: utf-8 -*-
"""Isra.ipynb

Automatically generated by Colaboratory.

Original file is located at
    https://colab.research.google.com/github/IsraelMelMon/CustomTrain-MobileNetV2/blob/master/Isra.ipynb

# Train your custom MobileNetV2

First, we give Google Colab permission to access Google Drive, and so, you must be signed-in to a [Google Drive](https://drive.google.com) account.
"""

from google.colab import drive

drive.mount('/content/drive/', force_remount=True)

"""We then move on to download the repository for an example. See the original Github repository [here](https://github.com/IsraelMelMon/CustomTrain-MobileNetV2)."""

cd drive/My Drive/

!git clone https://github.com/IsraelMelMon/CustomTrain-MobileNetV2.git

"""We go into the repository folder"""

cd CustomTrain-MobileNetV2

ls

"""The repository automatically downloads a folder in your Google Drive named "**CustomTrain-MobileNetV2/classes**" that contains the same number of folders as there are classes with images that are already classified."""

>/content/drive/My Drive/CustomTrain-MobileNetV2/classes

"""When adding your own images, each class should be a sub-folder contained in the **classes** folder.
The example given in the repository has two classes (although N different classes are supported) **IMPORTANT: Each image name should contain the class it corresponds to, for example:**
All the images in the folder, and so forth.
Let's run the initial configuration file.
```

Listing 9.17 Application of calibrate collaboration project.

```
"""

!ls

!python '/content/drive/My␣Drive/CustomTrain-MobileNetV2/
    auto\_setup.py'

"""By default, the training path is **/content/drive/My
    Drive/CustomTrain-MobileNetV2/classes/**, the learning
    rate is **1e-4**, batch size is **2**, training epochs
    are **50** and the output model name is **mbnv2.h5**.

To change this if you run the configuration file with other
    arguments as:
"""

"""Let's now run the training script:"""

!python '/content/drive/My␣Drive/CustomTrain-MobileNetV2/
    scripts/train.py'

"""If the training loss does not lower (it must change at
    least 1e-6 in a range of 20 epochs) then the training
    automatically stops and saves the last weights and the
    network before the loss was stunned as "**mbnv2.h5**".
    This file can now be used for predictive inference.

Specify that also multi-class classification may be done by
    configuring the Train.py file. And also the use of the
    ResNet50, ResNet101, etc supported in Tensorflow 2.x
"""

!pip freeze
```

Listing 9.17 (Continued.)

with the exclamation mark ! indicates execution on an Ubuntu bash. We use Keras==2.4.3, Keras-Preprocessing==1.1.2 and tensorflow==2.3.0. See listing 9.17.

Inference script
Having downloaded our previous h5 file, `mbnv2.h5`, we may now implement it for real-time video inferences.

9.5 End notes

In this chapter a complete application was implemented that allows to us recognize the pattern of different fruits, which is a solution to a typical industrial problem. A set of `Python` libraries was used under the paradigm of artificial intelligence and neural networks for decision-making in the classification.

References

Bhat S F, Alhashim I and Wonka P 2020 AdaBins: depth estimation using adaptive bins arXiv:2011.14141

Brown D C 1966 Decentering distortion of lenses *Photogramm. Eng.* **32** 444–62

Brown D C 1971 Close-range camera calibration *Photogramm. Eng.* **37** 855–66

Deng J, Guo J, Xue N and Zafeiriou S 2019 ArcFace: additive angular margin loss for deep face recognition arXiv: 1801.07698

Howard A G, Zhu M, Chen B, Kalenichenko D, Wang W, Weyand T, Andreetto M and Adam H 2017 MobileNets: efficient convolutional neural networks for mobile vision applications arXiv:1704.04861

Kleiman R and Page D 2019 AUC: a performance metric for multi-class machine learning models *Proc. of the 36th Int. Conf. on Machine Learning* vol 97*Proc. of Machine Learning Research(9–15 June 2019)*

Laga H, Jospin L V, Boussaid F and Bennamoun M 2020 A survey on deep learning techniques for stereo-based depth estimation *IEEE Trans. Pattern Anal. Mach. Intell.* https://doi.ieeecomputersociety.org/10.1109/TPAMI.2020.3032602

Luo X, Huang J-B, Szeliski R, Matzen K and Kopf J 2020 Consistent video depth estimation *ACM Trans. Graphics* **39** 71

Pozdnyakov D 2020 Fisheye lens distortion correction arXiv:2010.10295

Sandler M, Howard A, Zhu M, Zhmoginov A and Chen L-C 2018 MobileNetV2: inverted residuals and linear bottlenecks arXiv:1801.04381

Schroff F, Kalenichenko D and Philbin J 2015 FaceNet: a unified embedding for face recognition and clustering *IEEE Conf. on Computer Vision and Pattern Recognition* (Washington, DC: IEEE Computer Society) pp 815–23

Zhang K, Zhang Z, Li Z and Qiao Y 2016 Joint face detection and alignment using multitask cascaded convolutional networks *IEEE Signal Process. Lett.* **23** 1499–503

www.ingramcontent.com/pod-product-compliance
Ingram Content Group UK Ltd.
Pitfield, Milton Keynes, MK11 3LW, UK
UKHW051856100926
8556IPUK00005B/650